DEVELOPMENTS IN POLYMER DEGRADATION—6

CONTENTS OF VOLUMES 4 AND 5

Volume 4

Volume 5

DEVELOPMENTS IN POLYMER DEGRADATION—6

Edited by

N. GRASSIE

Chemistry Department, The University, Glasgow, UK

ELSEVIER APPLIED SCIENCE PUBLISHERS
LONDON and NEW YORK

ELSEVIER APPLIED SCIENCE PUBLISHERS LTD
Crown House, Linton Road, Barking, Essex IG11 8JU, England

Sole Distributor in the USA and Canada
ELSEVIER SCIENCE PUBLISHING CO., INC.
52 Vanderbilt Avenue, New York, NY 10017, USA

British Library Cataloguing in Publication Data

Developments in polymer degradation.—6
1. Polymers and polymerization—Deterioration
—Periodicals
I. Grassie, N.
668.9   QD381.8

ISBN-13: 978-94-010-8689-9   e-ISBN-13: 978-94-009-4940-9
DOI: 10.1007/978-94-009-4940-9

WITH 29 TABLES AND 109 ILLUSTRATIONS

© ELSEVIER APPLIED SCIENCE PUBLISHERS LTD 1985

Softcover reprint of the hardcover 1st edition 1985

PREFACE

The science of polymer degradation has many facets. Previous volumes have included representative topics with the purpose of building up a comprehensive picture of our current state of knowledge of the subject and its relevance to developing technology. In this present volume, the seven contributions represent four of the general areas of current interest, namely, products of degradation, degradative agencies, characterisation and the influence of the physical environment.

Detailed analyses of the products of degradation of polymers are of particular interest because of well justified current obsessions with toxicity, flammability and general safety of structural materials. Chapter 1, Dr Wright's review of the evolution of hydrogen fluoride from fluorine-containing elastomers, is therefore timely in view of the increasing application of fluorine-containing polymers as relatively stable structural materials.

The continual commercial pressure to apply polymers in more and more aggressive environments also continues to stimulate a great deal of current research and this is represented in Chapter 2 by Drs O'Donnell and Bowden on the radiation degradation of sulphones and in Chapter 3 by Professor Zaikov on the hydrolysis of a variety of polymers. The former is of additional interest in that it carries the subject from fundamental research to practical applications in microelectronic technology.

In the third category, Dr Henman in Chapter 4 describes a new technique for the analyses of oxidised structures in polyolefines while in Chapter 5 Professor Tüdős and his colleagues discuss recent work associated with the search for the labile structures which cause instability in poly(vinyl chloride).

It is becoming increasingly clear that the chemistry involved in polymer degradation processes can be profoundly influenced by the physical state of the polymer. In Chapter 6 Dr MacCallum shows, for example, how the rate of photo-oxidation may be influenced by the viscosity of the system while in Chapter 7 Dr Rapoport and Professor Zaikov demonstrate how stress may influence the rates of ageing processes.

All the contributions to this volume thus clearly have broad implications in polymer science and technology and should represent a useful addition to topics presented in previous volumes.

N. GRASSIE

CONTENTS

LIST OF CONTRIBUTORS

MURRAE J. BOWDEN

Bell Communications Research Inc., Murray Hill, New Jersey 07974, USA.

T. J. HENMAN

Cambridge Polymer Consultants, The Melbourn Science Park, Moat Lane, Melbourn, Royston, Herts SG8 6EJ, UK.

B. IVÁN

Central Research Institute for Chemistry, Hungarian Academy of Sciences, PO Box 17, Pusztaszeri ut 59-67, H-1525 Budapest, Hungary.

T. KELEN

Central Research Institute for Chemistry, Hungarian Academy of Sciences, PO Box 17, Pusztaszeri ut 59-67, H-1525 Budapest, Hungary.

J. P. KENNEDY

Institute of Polymer Science, University of Akron, Ohio, USA.

J. R. MACCALLUM

Department of Chemistry, University of St Andrews, The Purdie Building, St Andrews, Fife KY16 9ST, Scotland, UK.

James H. O'Donnell

Polymer and Radiation Group, Department of Chemistry, University of Queensland, Brisbane 4067, Australia.

N. Ya. Rapoport

Institute of Chemical Physics, Academy of Sciences of the USSR, Kosygin Street 4, Moscow 117334, USSR.

F. Tüdős

Central Research Institute for Chemistry, Hungarian Academy of Sciences, PO Box 17, Pusztaszeri ut 59-67, H-1525 Budapest, Hungary.

W. W. Wright

Materials and Structures Department, R178 Building, Royal Aircraft Establishment, Farnborough, Hants GU14 6TD, UK.

G. E. Zaikov

Institute of Chemical Physics, Academy of Sciences of the USSR, Kosygin Street 4, Moscow 117334, USSR.

Chapter 1

THE EVOLUTION OF HYDROGEN FLUORIDE FROM HYDROFLUORO-ELASTOMERS

W. W. Wright

Royal Aircraft Establishment, Farnborough, UK

SUMMARY

Hydrofluoro-elastomers comprise one of the most thermally stable elastomer systems currently available. Because of their chemical structure, there is the possibility of elimination of hydrogen fluoride from the polymer chains at elevated temperatures. This is analogous to the splitting out of hydrogen chloride from poly(vinyl chloride). If the hydrofluoro-elastomers are used as seals, or sealants, in contact with metals such as titanium, the presence of hydrogen fluoride may cause stress corrosion cracking. The available data on the evolution of hydrogen fluoride from hydrofluoro-elastomers are reviewed. It is shown that the type of formulation used in producing a practical rubber system is critical in this respect. Although substantial advances have been made in solving the technological aspects of the problem, there are still many unanswered questions relating to the mechanisms of the various processes involved.

1. INTRODUCTION

In his book on fluoropolymers published in 1972 Wall devotes only half a page to the thermal decomposition of fluorinated elastomers.[1] Coverage is restricted to copolymers of vinylidene fluoride with chlorotrifluoroethylene, or hexafluoropropene. He states that the thermal stability of these copolymers is of the same order as that of the homopolymers and that the

1

amount of hydrogen fluoride (HF) produced on thermal decomposition is relatively small. Perhaps because of this statement little attention has been paid to the possibility of HF evolution from these materials even though they comprise some of the most thermally stable elastomers currently available and, as such, would be expected to find their major use at elevated temperatures. What was the justification for Wall's statements and would they be accepted as still being true today?

2.   EARLY WORK UP TO 1972

The first paper which refers to the pyrolysis of these copolymers is by Wall himself.[2] In this the kinetics of decomposition of a copolymer of vinylidene fluoride with hexafluoropropene and of two copolymers of vinylidene fluoride with chlorotrifluoroethylene containing different proportions of comonomers were studied by thermogravimetry in vacuum. Samples were also pyrolysed under vacuum at different temperatures and four pyrolysis fractions isolated—a 'gas fraction' volatile at liquid nitrogen temperature, a 'condensed fraction' volatile at room temperature, a 'wax-like fraction' volatile at the temperature of pyrolysis and a 'residual fraction'. The first two fractions were analysed by mass spectrometry. The weight loss versus percentage conversion curves all showed maxima in the 20–40% conversion range indicating a random breakdown mechanism. Kinetic data derived from these maximum rates are summarised in Table 1 which also contains, for comparison purposes, similar results obtained by Wall and co-workers on the three homopolymers.[3,4]

These data confirm the similarity in thermal stability of the copolymers

TABLE 1

KINETIC DATA FOR THERMAL DEGRADATION OF FLUORINE-CONTAINING POLYMERS IN VACUUM[2-4]

Polymer composition	Activation energy $(kJ\,mol^{-1})$	Arrhenius factor (s^{-1})	Rate of weight loss at 350°C $(\%\,min^{-1})$
CF_2CH_2/CF_3CFCF_2 [70/30]	239	10^{13}	0·04
CF_2CH_2/CF_2CFCl [75/25]	256	10^{15}	0·06
CF_2CH_2/CF_2CFCl [64/36]	256	10^{15}	0·06
CF_2CH_2	202	10^{10}	0·02
CF_2CFCl	210	10^{12}	0·2
CF_3CFCF_2	238	10^{18}	76·8

TABLE 2

MASS SPECTRAL RESULTS FOR THERMAL DEGRADATION OF FLUORINE-CONTAINING POLYMERS IN VACUUM AT ABOUT $400\,°C^2$

Polymer composition	Total volatilised (%)	Light volatiles (as % of total volatilised)	HF (measured as SiF_4) (as % of light volatiles)
CF_2CH_2/CF_3CFCF_2 [70/30]	57·1	34·7 (19·8)	80·0 (15·8)
CF_2CH_2/CF_2CFCl [75/25]	85·4	23·1 (19·7)	80·3 (18·5)
CF_2CH_2/CF_2CFCl [64/36]	56·6	15·7 (8·9)	80·3 (12·6)

Figures in parentheses are the yields as a percentage of the original weights.

and poly(vinylidene fluoride) and poly(chlorotrifluoroethylene), but show that poly(hexafluoropropene) is much less stable.

The mass spectral results for pyrolysis at 400 °C are given in Table 2. These figures indicate that at 400 °C HF forms a significant fraction of the degradation products. Wright and his co-workers[5,6] studied the kinetics of breakdown of similar copolymers to Wall *et al.* by thermogravimetry in both vacuum and 300 mm oxygen pressure. Their data are summarised in Table 3. The figures for vacuum are in reasonable agreement with Wall's and the oxidative results show that the copolymers containing CF_2CFCl have a poorer stability in this atmosphere.

TABLE 3

KINETIC DATA FOR THERMAL DEGRADATION OF FLUORINE-CONTAINING COPOLYMERS IN VACUUM AND 300 mm OXYGEN PRESSURE[5,6]

Polymer composition	Atmosphere	Activation energy ($kJ\,mol^{-1}$)	Arrhenius factor (s^{-1})	Rate of weight loss at 350°C ($\%\,min^{-1}$)
CF_2CH_2/CF_3CFCF_2 [70/30]	Vacuum	193	10^{13}	0·04
CF_2CH_2/CF_3CFCF_2 [60/40]	Vacuum	189	10^{13}	0·05
CF_2CH_2/CF_2CFCl [67/33]	Vacuum	227	10^{16}	0·06
CF_2CH_2/CF_2CFCl [53/47]	Vacuum	210	10^{15}	0·12
CF_2CH_2/CF_2CFCl [19/81]	Vacuum	286	10^{21}	0·18
CF_2CH_2/CF_3CFCF_2 [70/30]	Oxygen	(176)	(10^{11})	0·03
CF_2CH_2/CF_3CFCF_2 [60/40]	Oxygen	—	—	~0·07
CF_2CH_2/CF_2CFCl [67/33]	Oxygen	155	10^{11}	0·69
CF_2CH_2/CF_2CFCl [53/47]	Oxygen	151	10^{11}	1·30
CF_2CH_2/CF_2CFCl [19/81]	Oxygen	164	10^{12}	0·67

Degteva *et al.*[7-9] studied the degradation of the [53/47] copolymer of vinylidene fluoride and chlorotrifluoroethylene in vacuum and 360 mm oxygen pressure measuring the overall activation energy for degradation and also the activation energies for evolution of HCl and HF. In vacuum, evolution of hydrogen halides commenced in the temperature range 250–300 °C and was in the ratio 2:1 for HCl:HF. The activation energies for production of HCl and HF were 122 and 143 kJ mol^{-1}, respectively. At temperatures greater than 300 °C the position changed considerably and about three times as much HF was evolved as HCl. The activation energy for formation of HCl also increased to 193 kJ mol^{-1} and that for formation of HF to 231 kJ mol^{-1}. Degteva claimed that some fluorine was also produced and that the results changed somewhat if the experiments were carried out in platinum rather than in glass. In oxygen, at temperatures greater than 300 °C and using a platinum reaction vessel, very similar activation energies were obtained for overall degradation (as measured by oxygen absorption) and HCl and HF evolution—all were in the range 109–113 kJ mol^{-1}.

Degteva and Kuzminskii studied the effect of practical mix components

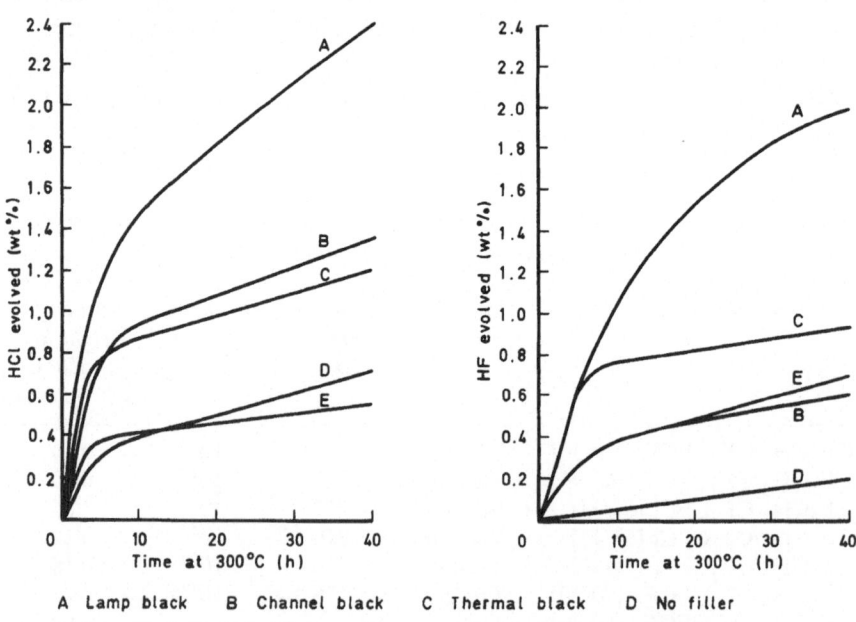

A Lamp black   B Channel black   C Thermal black   D No filler
E Furnace black (All fillers at 50 parts per 100 of copolymer loading)

FIG. 1.   Comparison of the effects of carbon black fillers on the evolution of HCl and HF from a vinylidene fluoride/chlorotrifluoroethylene copolymer at 300 °C in vacuum.[10]

upon the evolution of hydrogen halides from the [53/47] copolymer of vinylidene fluoride and chlorotrifluoroethylene in vacuum at 250–300 °C and found some very interesting results.[10] The addition of carbon blacks accelerated the evolution of HCl and HF, but not to the same degree or in the same order (Fig. 1). For HCl the order of evolution was lamp black > channel black ≈ thermal black > no filler ≈ furnace black, whereas for HF the order was lamp black > thermal black > furnace black ≈ channel black > no filler. For minimum evolution of hydrogen halides, furnace black would therefore be the best filler to use. A number of metallic oxides, Al_2O_3, CaO, Co_2O_3, Fe_2O_3, MgO, PbO and ZnO, which would be regarded as halide acceptors, were also tested and the results are summarised in Fig. 2. The data for Fe_2O_3 and CaO are not included, the evolution of hydrogen halides being too rapid with the former (30 % HF and 5·4 % HCl in 2 h at temperature) and too slow with the latter to fit sensibly on the scales used. The majority of the oxides caused an increase in the rate of evolution of both the hydrogen halides, although not to the same degree. Zinc oxide accelerated the evolution of HF, but reduced the evolution of HCl. Only PbO and CaO reduced the liberation of both compounds; the latter to a much greater degree. In the presence of

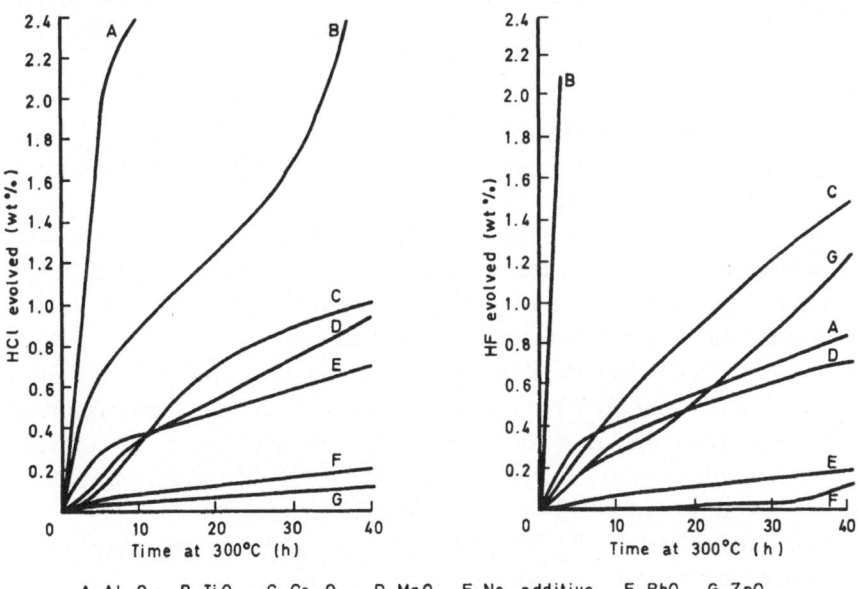

A Al_2O_3 B TiO_2 C Co_2O_3 D MgO   E No additive   F PbO   G ZnO

FIG. 2. Comparison of the effects of metallic oxides on the evolution of HCl and HF from a vinylidene fluoride/chlorotrifluoroethylene copolymer at 300 °C in vacuum.[10]

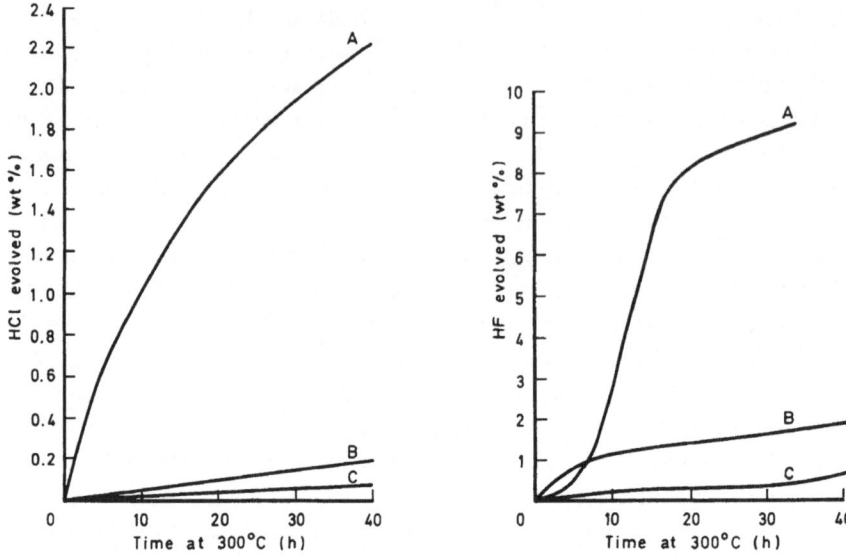

A Graphite (50 parts)   B Graphite (50 parts) + CaO (10 parts)
C Graphite (50 parts) + CaO (10 parts) + zinc diethyldithiocarbamate (1 part)

FIG. 3.   Comparison of the effects of various additives on the evolution of HCl and HF from a vinylidene fluoride/chlorotrifluoroethylene copolymer at 300°C in vacuum.[10]

graphite, however, CaO was by no means as effective (Fig. 3). Some improvement was attained by the addition of 1 part of zinc diethyldithiocarbamate to the mixture.

All the results listed so far have been for mixtures of the copolymer with other materials. In practice, of course, the copolymer is crosslinked (vulcanised) using either a diamine, or a peroxide. The effect of this on hydrogen halide evolution is summarised in Table 4.

It can be seen that crosslinking destabilised the copolymer, the most drastic effect being observed for HF evolution from a diamine-vulcanised material, where the rate was increased about 250 times. In contrast, peroxide vulcanisation increased the rate of HF liberation only about 3 times and that of HCl evolution about 17 times.

No further work on this particular copolymer is reported in the literature. This is understandable from a commercial point of view as the market has become dominated by copolymers of vinylidene fluoride with other fluorine-containing (but not chlorine-containing) monomers. It is surprising, however, that the material has not received more attention

TABLE 4

EFFECT OF CROSSLINKING ON HYDROGEN HALIDE EVOLUTION FROM A VINYLIDENE FLUORIDE/CHLOROTRIFLUOROETHYLENE COPOLYMER AT 300 °C IN A VACUUM[10]

Material studied	HCl evolved (weight %) after heating for			HF evolved (weight %) after heating for		
	4 h	18 h	28 h	4 h	18 h	28 h
Copolymer	0·26	0·53	0·68	0·03	0·12	0·15
Copolymer crosslinked with a diamine	0·96	7·67	8·23	10·07	37·60	38·80
Copolymer crosslinked with a peroxide	0·15	3·04	12·29	0·11	0·32	0·50

academically, because of the interest and complexity of the effects observed and the lack of an overall explanation for them.

Only two further papers were published in this area up to 1972 and these were both on copolymers of vinylidene fluoride and hexafluoropropene. Degteva et al.[11] studied degradation in vacuum over the temperature range 250–400 °C using the same techniques as they had employed for the copolymer containing chlorotrifluoroethylene. After 40 h at 320 °C only 0·1 % by weight of HF had evolved; the total weight loss in this time was about 1 %. If the temperature was raised to 360–400 °C the copolymer was almost completely destroyed and the HF yield increased reaching about 2 % after 5 h at 400 °C. Oksentevich and Pravednikov[12] also investigated the same copolymer at 360–415 °C in vacuum. They derived an overall activation energy of degradation (from maximum rates of weight loss) of 248 kJ mol^{-1}. The quantity of HF liberated did not exceed 2 % by weight.

The data available in 1972, therefore, supported Walls' statement that the amount of HF obtained on thermal decomposition of these fluorine-containing copolymers was relatively small. It should be noted, however, that most of the experiments had been carried out on the 'pure' copolymers in vacuum and there were already indications that compounding could have quite major effects.

3. STUDIES FROM 1972 TO DATE

Brown and Wall[13] investigated the high temperature ageing of fluoropolymers and included a vinylidene fluoride/hexafluoropropene copolymer. After 87 h at 225 °C in vacuum there was less than 1 % weight loss. At 305 °C

TABLE 5

COMPARISON OF WEIGHT LOSS AND HF YIELD FOR COPOLYMERS OF VINYLIDENE FLUORIDE AND HEXAFLUOROPROPENE

Copolymer	Atmosphere	Temperature °C for				Final HF yield
		Initial weight loss	1% weight loss	Initial HF yield	1% HF yield	
CF_2CH_2/CF_3CFCF_2 [61/39]	Nitrogen	270	350	136	457	12·9
	Air	340	395	143	431	54·2
CF_2CH_2/CF_3CFCF_2, higher molecular weight [61/39]	Nitrogen	350	410	141	430	13·2
	Air	380	420	195	417	54·7
CF_2CH_2/CF_3CFCF_2, crosslinked [61/39]	Nitrogen	170	320	188	361	28·5
	Air	200	360	220	383	56·2
CF_2CH_2/CF_3CFCF_2 [49/51]	Nitrogen	400	440	271	442	13·5
	Air	370	425	259	420	46·5

after the same time, the weight loss had risen to 60% and 0·3% HF (measured as SiF_4) was detected. In air at 225°C, however, considerable crosslinking occurred and a gel fraction of 0·35 was obtained in 15 h.

Knight and Wright[14] embarked on a series of studies specifically aimed at determining the amount of HF evolution from hydrofluoro and perfluoro polymers and copolymers in nitrogen and air. They utilised a fluoride ion-specific electrode and a flow system so that the evolution of HF could be monitored continuously, either as a function of rising temperature, or as a function of time at constant temperature. The overall weight loss under similar experimental conditions was determined separately. Their results for three different copolymers of vinylidene fluoride and hexafluoropropene obtained at a rate of temperature rise of 2°C min^{-1} are summarised in Table 5, which also includes data for one of the copolymers, which had been crosslinked using N,N-dicinnamylidine-1,6-hexane diamine. On the basis of the weight loss figures in nitrogen and in air the crosslinked material was the least stable, followed by the non-crosslinked analogue and then by the copolymers of higher molecular weight and slightly different formulation, the last containing a smaller proportion of vinylidene fluoride. The HF yield results were not so clear cut. Evolution of HF apparently commenced at much lower temperatures than weight loss, but this is related to the sensitivity of the technique used for detection of HF, which easily monitored changes at concentrations of the order of 10^{-5} mol litre^{-1} fluoride ion. The temperatures for 1% weight loss and 1% HF yield were generally of the same order and it is noticeable that the temperatures for 1% HF yield from the crosslinked material were well below those for the non-crosslinked copolymers. Comparison of the weight loss and HF yield curves showed that at least half of the HF finally evolved must have arisen from secondary breakdown of volatile (at temperature of pyrolysis) degradation products. This was true for experiments in nitrogen and in air.

Following this indication that crosslinking with a diamine destabilised a vinylidene fluoride/hexafluoropropene copolymer towards dehydro-fluorination, Knight and Wright studied[15] a number of such copolymers in gumstock and fully compounded form and in addition to dynamic experiments in nitrogen and air carried out isothermal measurements in air in the temperature range 200–275°C. The copolymers were fairly similar in composition, the hydrogen content only varying from 1·7 to 2·3 per cent and the fluorine content from 63·8 to 66·1 per cent. Compounding consisted of crosslinking with a diamine, or a hydroquinone derivative and the addition of 15–30 parts medium thermal black and 5-15 parts

magnesium oxide. In some cases 2–6 parts calcium hydroxide were also
added. (All parts were per hundred parts of copolymer.) The compounded
material was press-cured and in some cases post-cured as well. Confirming
the earlier results (Table 5), higher yields of HF were obtained in air than in
nitrogen. In experiments with the temperature rising at $2\,°C\,min^{-1}$, by
500 °C the HF yield in air varied from 22 to 48 % of that theoretically
possible, whereas in nitrogen the figures were 5 to 25 %. A comparison with
weight loss results showed that for all the samples in air the process leading
to HF evolution was the major breakdown route in the initial stages of
degradation. An important practical point was that the HF yield from
some formulations containing as much as 15 per cent magnesium oxide as
acid acceptor was greater than the yield from the uncompounded, non-
crosslinked copolymer. This was confirmed in isothermal experiments,
where it could be seen that the rate of HF evolution fell off rapidly with time
of heating for the virgin copolymers, whereas the rate of production
remained essentially constant over the timescale of the experiments for the
compounded materials. That is, in the latter case, apparent zero order
kinetics were obeyed. The activation energies for HF production derived
from these data lay in the range $92–134\,kJ\,mol^{-1}$, the majority of results
being between 117 and $134\,kJ\,mol^{-1}$. One particular copolymer, which was
examined in pure form and also compounded, press-cured and post-cured,
gave the same value for the activation energy in all cases, i.e. the various
fabrication procedures had not affected the initial rate of HF yield. In all
cases, except where calcium hydroxide was included in the formulation,
HF liberation was observed from the very start of the experiments. With
calcium hydroxide present an induction period occurred. The results
showed that HF evolution could be a problem in practical applications if
some particular formulations were used at temperatures above 150 °C.
They also demonstrated that with the particular copolymers studied,
compounding was of more importance with respect to HF yield than the
variations in the composition of the copolymers themselves. Accordingly,
one particular system was examined in detail to determine the effect of each
individual component upon HF production.

The system chosen for study was basically a terpolymer of vinylidene
fluoride, hexafluoropropene and tetrafluoroethylene,[16] containing 1·8 %
hydrogen and 65·0 % fluorine. The normal formulation of this elastomer
consists of 100 parts by weight terpolymer, 20 parts medium thermal black,
15 parts magnesium oxide and 3 parts N,N-dicinnamylidene-1,6-hexane-
diamine (curing agent) and the results for this formulation were used as a
norm for comparison of all the other formulations tested. Figure 4 shows

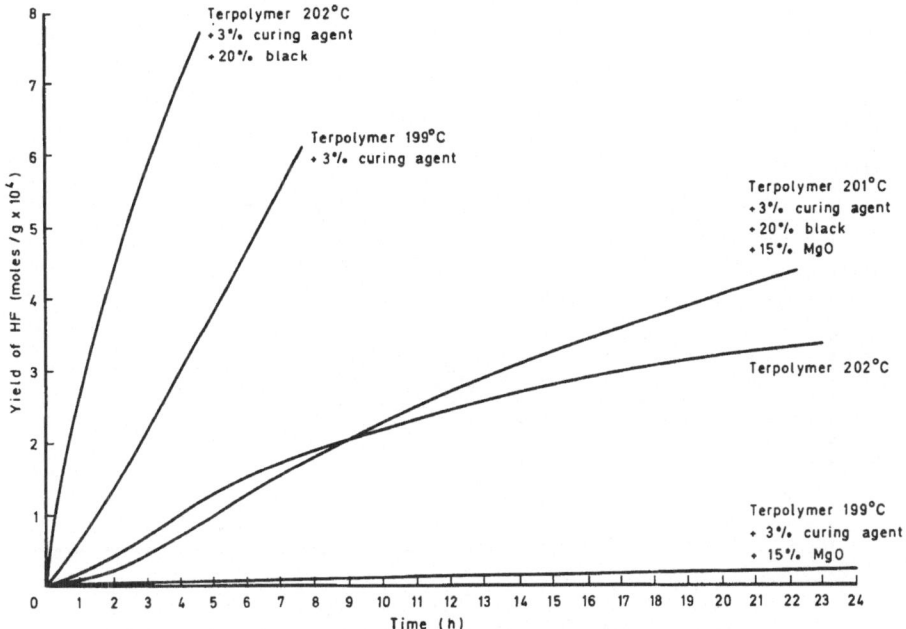

FIG. 4. Effect of change in formulation on HF yield at 200 °C.[16]

the effect on HF yield at approximately 200 °C of omitting one or more of the components listed in the above formulation. The same relative behaviour occurred at temperatures at least up to 275 °C. It can be seen that crosslinking the terpolymer with a diamine markedly increased the amount of HF evolved, as had been observed by Degteva *et al.*[10] for the copolymer of vinylidene fluoride and chlorotrifluoroethylene. Addition of medium thermal black caused further destabilisation. If magnesium oxide was then also added the stability was almost restored to that of the unmodified terpolymer, although in the long term more HF was evolved from the compounded material. In the absence of black, the magnesium oxide was much more effective, the yield of HF being reduced by at least a factor of ten below that for the terpolymer alone.

As the presence of the thermal black was having such an obvious effect upon HF yield a number of other types of black were tested. All formulations contained 15% magnesium oxide and were cured with diamine. The results at approximately 275 °C are shown in Fig. 5. The data demonstrate that in contrast to thermal black, channel and furnace black were relatively inactive. This is in agreement with Degteva's findings[10] for HF evolution from a copolymer of vinylidene fluoride and chlorotrifluoro-ethylene.

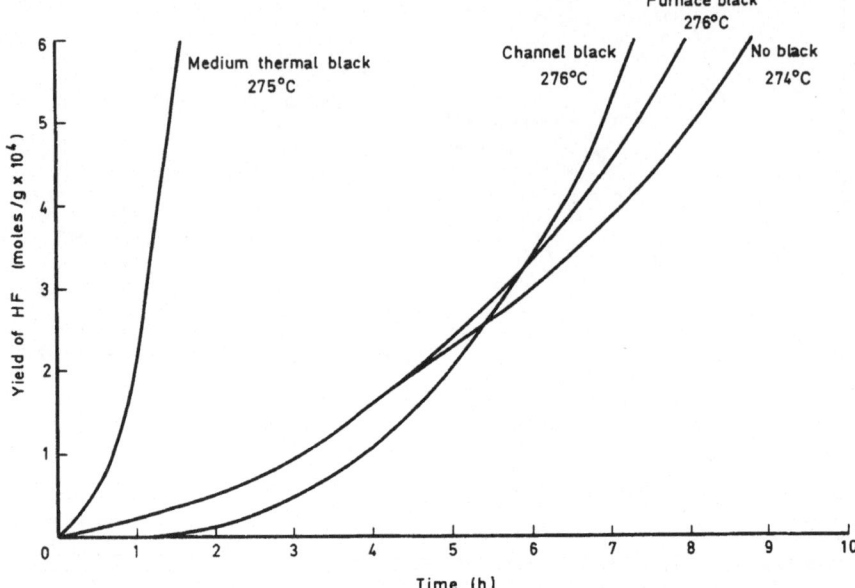

FIG. 5.   Effect of type of carbon black on HF yield at 275 °C.[16]

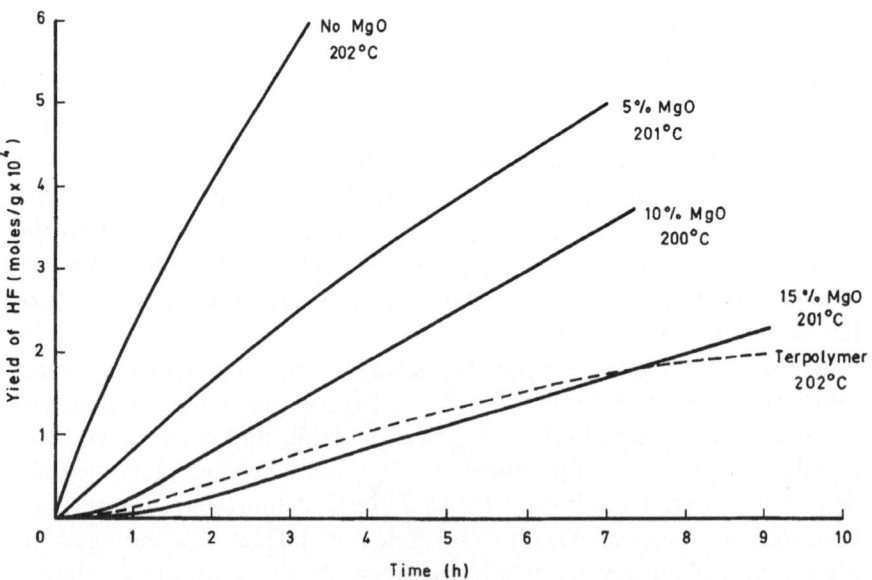

FIG. 6.   Effect of amount of magnesium oxide on HF yield at 200 °C.[16]

The effect of varying the magnesium oxide content was also investigated. In this case all formulations contained 20% medium thermal black and were cured with diamine. Figure 6 illustrates the variation in HF yield at 200°C with magnesium oxide content. The addition of MgO did not prevent HF elimination, but reduced its rate of evolution, a 15% addition giving a result comparable with that for the terpolymer alone. The reduction in rate did not have a linear dependence upon MgO concentration, as can be seen in Fig. 7. Similar results were obtained at temperatures up to 275°C.

The efficiencies of a number of basic oxides, hydroxides and carbonates as HF acceptors were compared with that of magnesium oxide in the temperature range 200–275°C. The results are summarised in Table 6. All the formulations contained 20% thermal black and were cured with diamine. Only two of the compounds tested, namely TiO_2 and K_2CO_3, were less efficient than MgO. Of the remainder, CaO, $Mg(OH)_2$, $Ca(OH)_2$, $CaCO_3$, $MgCO_3$ and Li_2CO_3 were manifestly superior. The relative efficiencies depended to a certain extent upon the temperature of measurement. At 225°C, Li_2CO_3 and $MgCO_3$ were the most effective,

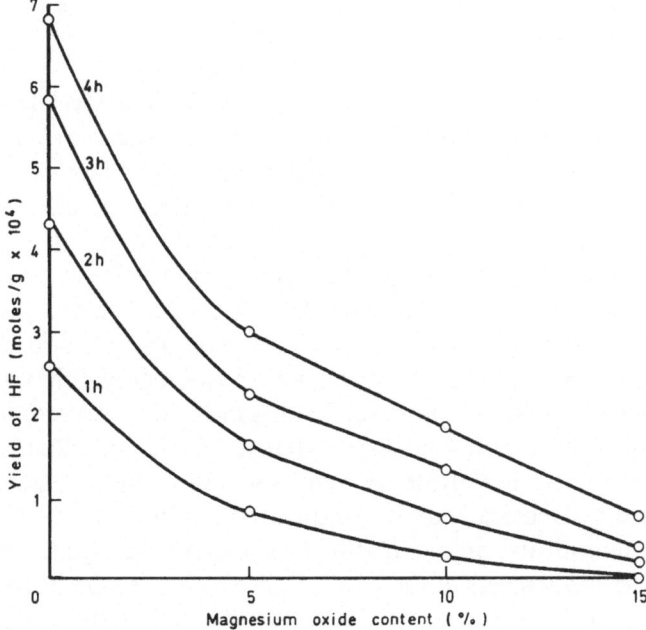

FIG. 7. Dependence of HF yield on magnesium oxide content as a function of time.

TABLE 6

COMPARISON OF BASIC OXIDES, HYDROXIDES AND CARBONATES AS HF ACCEPTORS

Additive (15% by weight)	Relative efficiency as HF acceptor at			
	200°C	225°C	250°C	275°C
TiO_2	0·5	0·2	0·2	0·2
MgO	1·0	1·0	1·0	1·0
BeO	1·7	1·6	2·2	1·5
Al_2O_3	2·9	1·8	1·7	1·5
B_2O_3	3·7	2·7	0·8	2·1
Li_2O	—	3·4	3·4	1·5
CaO	—	38·8	>48·0	25·6
$Al(OH)_3$	1·2	2·2	2·7	2·2
LiOH	—	32·2	10·6	6·9
$Mg(OH)_2$	—	31·6	44·3	37·0
$Ca(OH)_2$	—	54·7	>64·0	42·5
K_2CO_3	0·6	0·7	>0·1	0
Na_2CO_3	—	17·6	3·4	2·6
$CaCO_3$	—	87·4	56·6	41·3
$MgCO_3$	—	115·3	25·5	12·2
Li_2CO_3	—	137·6	32·4	21·4

whereas at higher temperatures $Ca(OH)_2$, $CaCO_3$, and $Mg(OH)_2$ reduced HF evolution by the greatest amounts. These differences are illustrated in Fig. 8. In their study[10] of the unvulcanised copolymer of vinylidene fluoride and chlorotrifluoroethylene at 300°C, Degteva and Kuzminskii used metallic oxides as the sole additive. They found an order of relative efficiency of $CaO > PbO > MgO > Al_2O_3 > ZnO > Co_2O_3 > TiO_2$. The effectiveness of CaO was, however, greatly reduced in the presence of graphite (see Fig. 3). Degteva and Kuzminskii also claimed that combinations of CaO with sodium nitrite, or with zinc diethyldithiocarbamate, were extremely good acceptors for hydrogen halides. Accordingly, these combinations were tested in the vinylidene fluoride/hexafluoropropene/tetrafluoroethylene terpolymer system, as was PbO, the second best of the oxides used by Degteva et al. In contrast to their findings, the addition of either sodium nitrite, or zinc diethyldithiocarbamate together with CaO reduced the effectiveness of the latter. Furthermore, PbO was found to be completely ineffective as an HF acceptor. The addition of a polymeric antioxidant, however, was advantageous in the case of MgO, but had relatively little effect with $Mg(OH)_2$.

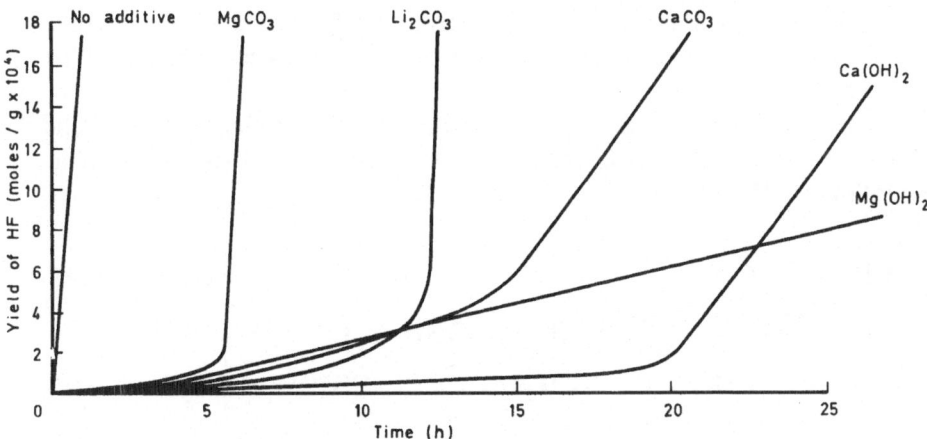

FIG. 8. Comparison of HF evolution at 275 °C in the presence of various additives.

The overall energies for HF production were calculated from the initial rates of HF yield measured at different temperatures and the results available are summarised in Table 7.

The most interesting feature of these results is the relatively small variation observed in activation energy from that determined for the terpolymer itself, that is, despite crosslinking, addition of medium thermal

TABLE 7

OVERALL ACTIVATION ENERGIES FOR HF YIELD FROM DIFFERENT FORMULATIONS

Formulation	Activation energy $(kJ\,mol^{-1})$
Terpolymer	102
Terpolymer cured with diamine	111
Terpolymer cured with diamine + 15% MgO	105
Terpolymer cured with diamine + 20% MT black	90
Terpolymer cured with diamine + 20% MT black + 5% MgO	102
Terpolymer cured with diamine + 20% MT black + 10% MgO	98
Terpolymer cured with diamine + 20% MT black + 15% MgO	87
Terpolymer cured with diamine + 20% MT black + 15% TiO_2	121
Terpolymer cured with diamine + 20% MT black + 15% Al_2O_3	122
Terpolymer cured with diamine + 20% MT black + 15% BeO	113
Terpolymer cured with diamine + 20% MT black + 15% B_2O_3	99
Terpolymer cured with diamine + 20% MT black + 15% $Al(OH)_3$	99
Terpolymer cured with diamine + 20% MT black + 15% K_2CO_3	95

black and the acid acceptors cited, approximately the same rate of change
of HF evolution with temperature was observed. This could be useful for
predictive purposes.

Are there any satisfactory explanations for the results observed? The
cure of a hydrofluoro-elastomer with a diamine is said to take place as
follows:[17,18]

$$
\begin{array}{cccccc}
| & | & | & | & | & | \\
CH_2 & CH & CH_2 & CH_2 & CH_2 & CH_2 \\
| & \| & | & | & | & | \\
CF_2 \xrightarrow{-HF} & CF \xrightarrow{H_2NRNH_2} & CF-NHRNH-CF & \xrightarrow{-HF} & C=NRN=C \\
| & | & | & | & | & | \\
CH_2 & CH_2 & CH_2 & CH_2 & CH_2 & CH_2 \\
| & | & | & | & | & | \\
\end{array}
$$

(A)   (B)   (C)

(A) Elimination of HF from the vinylidene fluoride segments of the
     chain to produce double bonds
(B) Addition of the diamine to these double bonds to form crosslinks
(C) Further elimination of HF from the crosslinks to yield more double
     bonds

The increase in HF yield, when the terpolymer is crosslinked with a
diamine is, therefore, understandable, especially as the double bonds
produced in the crosslinks would tend to favour elimination of HF from the
next units in the chain.

The carbon blacks used differed in particle size, oxygen content and pH
value (Table 8).

It is not immediately apparent from these figures, why medium thermal
black in the presence of MgO should have accelerated HF evolution,
whereas furnace and channel black did not. The problem is compounded
because, when used with $Ca(OH)_2$, medium thermal black proved to be
preferable to furnace black.

The relative efficiencies, at different temperatures, of the basic oxides,

TABLE 8

PROPERTIES OF DIFFERENT TYPES OF CARBON BLACK

Black type	Particle size (nm)	Oxygen content (%)	pH
Medium thermal	300	0	7·0
Furnace	70	0·22	9·5
Channel	32	3·6	4·5

hydroxides and carbonates are also difficult to explain. All samples were ground and sieved to give the same particle size distribution and the efficiencies quoted take account of the theoretical HF absorption per unit weight. The effect of the fluorides to which the various compounds would be converted during HF absorption upon further HF evolution was determined and shown to be relatively insignificant. It should be noted that the compounds do not function solely as HF acceptors, but also play a vital part in the curing reaction and hence it must be ascertained that in their presence an adequate crosslink density is attained. This is certainly the case for the preferred additives listed above. The thermodynamics of the reactions of these materials with HF were analysed, but gave no indication of why they should have such vastly different efficiencies as HF acceptors.

Kiroshko and Grimblat have studied[19] the dehydrofluorination of the unvulcanised copolymers of vinylidene fluoride with hexafluoropropene, or perfluoromethylvinylether over the temperature range 200–350 °C in air and argon. Stability up to 250 °C was good, only 0·06–0·08 % HF being evolved during 8 h at that temperature in air and 0·02–0·03 % in argon. Even at 350 °C the maximum amount of HF produced was 0·35 %. The activation energies for HF evolution in air were 87 and 77 kJ mol^{-1} for the copolymers containing hexafluoropropene and perfluoromethylvinylether, respectively.

The most recent work[20,21] has examined some of the newer fluorine-containing elastomers. These include copolymers of vinylidene fluoride and hexafluoropropene crosslinked with either diamines or peroxides, and copolymers of tetrafluoroethylene with propene (also cured with organic peroxides), or with perfluoromethylvinylether. The behaviour of the copolymer of vinylidene fluoride and hexafluoropropene crosslinked with diamine was very similar to that already cited for material of this type, that is, after an initial period the yield of HF was much greater from the compounded, crosslinked material than from the copolymer itself (Fig. 9). Quite different behaviour was, however, observed for the copolymer of tetrafluoroethylene and propene crosslinked with a peroxide, where the stability of the compounded material was much better than that of the virgin material (Fig. 9). The results, in fact, were vastly superior to those for a diamine crosslinked system containing the best of the HF acceptors investigated. This could only be a function of the peroxide cure as the stabilities of the copolymers themselves were very similar. (Degteva et al.[10] had observed similar effects in the crosslinking of the copolymer of vinylidene fluoride and chlorotrifluoroethylene with a diamine and a peroxide.) Up to about 50 h at temperature, the results were comparable

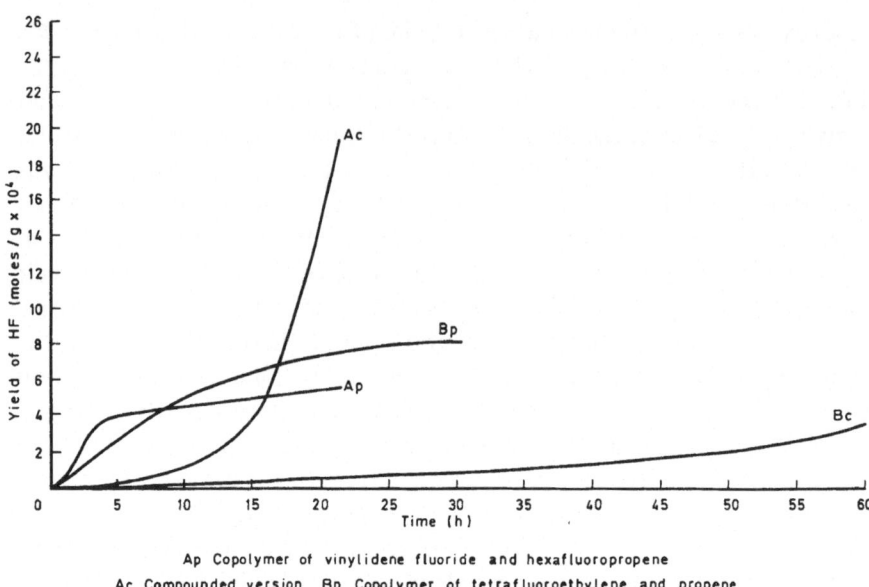

Ap Copolymer of vinylidene fluoride and hexafluoropropene
Ac Compounded version   Bp Copolymer of tetrafluoroethylene and propene
Bc Compounded version

FIG. 9.   Comparison of HF yields from two copolymers at 275 °C.

with those obtained for a compounded sample of the copolymer of tetrafluoroethylene and perfluoromethylvinylether (in reality this is a terpolymer containing a small amount of a third, undisclosed monomer to facilitate crosslinking. Possible monomers are $CF_2{=}CFO(CF_2)_4CN$, $CF_2{=}CFO(CF_2)_4COOCH_3$, $CF_2{=}CFOCF_2CF(CF_3)OC_6F_5$, $CF_2{=}CFO(CF_2)_3OC_6F_5$, which being a fully fluorinated material would be expected to give a very low HF yield. It is only at times greater than 50 h that the expected superiority becomes evident.

4.   CONCLUSIONS

In contrast to the conclusion of Wall, based upon results derived mainly in inert atmospheres and on pure copolymers, that the amount of HF evolved from hydrofluoro-elastomers upon thermal degradation is relatively small, more recent studies have demonstrated that in air and using compounded, crosslinked material the amount can be of practical significance. Stress-corrosion cracking of metals such as titanium and stainless steel has been observed at temperatures as low as 160 °C, when fluorine-containing seals and sealants have been used in contact with them. The data obtained show clearly the main steps to be taken to minimise the possibility of this happening. The first and most important is to use a peroxide-cured rather than a diamine-cured system. Within diamine-cured systems advantages accrue by using furnace, or channel black, rather than medium thermal black and acid acceptors such as CaO, $Mg(OH)_2$, $Ca(OH)_2$, $CaCO_3$, $MgCO_3$ and Li_2CO_3 rather than MgO. The use of these materials to reduce HF elimination must, of course, be balanced against their possible effects upon other desired properties. Although the work done enables these technological recommendations to be made, relatively little light has been shed upon the reasons for the observed differences in behaviour. This could prove a fruitful area for academic study.

REFERENCES

1. WALL, L. A. (Ed.), *Fluoropolymers*, New York, Wiley-Interscience, 1972, p. 414.
2. WALL, L. A. and STRAUS, S., *J. Research Nat. Bur. Stand.*, **65A** (1961), 227.
3. MADORSKY, S. L. and STRAUS, S., *J. Research Nat. Bur. Stand.*, **63A** (1959), 261.
4. STRAUS, S. and WALL, L. A., *Soc. Plast. Eng. Trans.* **4** (1964), 56.
5. COX, J. M., WRIGHT, B. A. and WRIGHT, W. W., *J. Applied Polym. Sci.*, **8** (1964), 2935.
6. COX, J. M., WRIGHT, B. A. and WRIGHT, W. W., *J. Applied Polym. Sci.*, **8** (1964), 2951.
7. DEGTEVA, T. G. and KUZMINSKII, A. S., *Vysokomol. Soed.*, **5** (1963), 1417.
8. DEGTEVA, T. G., SEDOVA, I. M. and KUZMINSKII, A. S., *Vysokomol. Soed.*, **5** (1963), 378.
9. DEGTEVA, T. G., SEDOVA, I. M. and KUZMINSKII, A. S., *Vysokomol. Soed.*, **5** (1963), 1485.
10. DEGTEVA, T. G. and KUZMINSKII, A. S., *Soviet Rubber Technology*, **23** (1964), 13.
11. DEGTEVA, T. G., SEDOVA, I. M., KHAMIDOV, K. A. and KUZMINSKII, A. S., *Vysokomol. Soed.*, **7** (1965), 1198.
12. OKSENTEVICH, L. A. and PRAVEDNIKOV, A. N., *Vysokomol. Soed. Krat. Soobshch*, **10B** (1968), 49.
13. BROWN, D. W. and WALL, L. A., *J. Polym. Sci., Polym. Chem. Ed.*, **10** (1972), 2967.
14. KNIGHT, G. J. and WRIGHT, W. W., *J. Applied Polym. Sci.*, **16** (1972), 683.
15. KNIGHT, G. J. and WRIGHT, W. W., *Br. Polym. J.*, **5** (1973), 395.
16. WRIGHT, W. W., *Br. Polym. J.*, **6** (1974), 147.
17. SMITH, J. F., *Rubber World*, **142** (1960), 102.
18. SMITH, J. F. and PERKINS, C. T., *J. Applied Polym. Sci.*, **5** (1961), 460.
19. KIROSHKO, P. B. and GRIMBLAT, M. P., *Kauch i Rezina*, **3** (1974), 6.
20. KNIGHT, G. J. and WRIGHT, W. W., *Thermochimica Acta*, **60** (1983), 187.
21. KNIGHT, G. J. and WRIGHT, W. W., *Polym. Degrad. Stab.*, **4** (1982), 465.

Chapter 2

RADIATION DEGRADATION OF POLY(OLEFIN SULPHONE)S—FUNDAMENTAL RESEARCH TO PRACTICAL APPLICATIONS

MURRAE J. BOWDEN

Bell Communications Research Inc., Murray Hill, New Jersey, USA

and

JAMES H. O'DONNELL

*Polymer and Radiation Group, Department of Chemistry,
University of Queensland, Brisbane, Australia*

SUMMARY

The degradation of poly(olefin sulphone)s by high energy radiation, particularly by γ-rays and electron beams, provides a fascinating example of the application of fundamental science to high technology industry. Scientific interest in the radiation degradation of these polymers was first aroused by the discovery that they underwent highly specific bond scission in the backbone chain as the primary result of absorption of high energy radiation and in fact they were the first polymers in which such an effect had been demonstrated. This conclusion was initially based mainly on evidence from electron spin resonance spectroscopy and was subsequently verified by studies of molecular weight changes. These studies showed that the poly(olefin sulphone)s not only degraded by main chain scission but were also among the most radiation-sensitive polymers known.

The extremely high sensitivity of poly(olefin sulphone)s to radiation-induced main-chain scission has found application in the field of microelectronics. Electron beam writing on poly(olefin sulphone) films is used to produce lithographic masks for the manufacture of integrated circuits on silicon wafers. Poly(1-butene sulphone) (PBS) is currently used in the production of a substantial proportion of the masks for the industry.

It is the purpose of this paper to review the fundamental aspects of the radiation degradation of poly(olefin sulphone)s along with the practical applications to high technology stemming from the fundamental science.

1.   INTRODUCTION

The degradation of poly(olefin sulphone)s by high energy radiation, particularly by γ-rays and electron beams, provides a fascinating example of the application of fundamental science to high technology industry. Scientific interest in the radiation degradation of these polymers was first aroused by the discovery that they underwent highly specific bond scission in the backbone chain as the primary result of absorption of high energy radiation and in fact they were the first polymers in which such an effect had been demonstrated. This conclusion was initially based mainly on evidence from electron spin resonance spectroscopy and was subsequently verified by studies of molecular weight changes. These studies showed that the poly(olefin sulphone)s not only degraded by main chain scission but were also among the most radiation-sensitive polymers known. The major radiolysis products were SO_2 and olefin, the yields of which increased exponentially with increasing irradiation temperature. These results were interpreted in terms of the ceiling temperature phenomenon which had been investigated previously in the preparation of these polymers.

The extremely high sensitivity of poly(olefin sulphone)s to radiation-induced main-chain scission has found application in the field of microelectronics. Electron beam writing on poly(olefin sulphone) films is used to produce lithographic masks for the manufacture of integrated circuits on silicon wafers. Poly(1-butene sulphone) (PBS) is currently used in the production of a substantial proportion of the masks for the industry.

It is the purpose of this paper to review the fundamental aspects of the radiation degradation of poly(olefin sulphone)s along with the practical applications to high technology stemming from fundamental science.

2.   POLYMER PREPARATION AND STRUCTURE

2.1. Alternating Poly(Olefin Sulphone)s

Poly(olefin sulphone)s, or poly(sulphonylalkylene)s, are copolymers containing sulphur dioxide and two carbon-atom units in the chain, i.e.

$-SO_2-CHR-CH_2-$, or more generally $-SO_2-CR_2-CR_2-$. They are usually prepared by the free radical copolymerisation of SO_2 and an olefin, or olefinic compound, in the liquid state, either in bulk or in solution, according to eqn (1).[1-3]

$$CH_2=\underset{R}{\overset{H}{C}} + SO_2 \longrightarrow \left[\underset{\underset{H}{|} \quad \underset{R}{|} \quad \underset{O}{||}}{\overset{\overset{H}{|} \quad \overset{H}{|} \quad \overset{O}{||}}{C-C-S}} \right]_n \qquad (1)$$

I

They can also be prepared by copolymerisation in the gas phase,[4,5] and by oxidation of polysulphides,[6-8] prepared by ionic polymerisation of alkylene sulphides.

A large number of olefins, M, have been copolymerised with sulphur dioxide, S, to give copolymers with an average composition ratio, $\bar{n} = M/S = 1.0$. Moreover, n.m.r. studies have shown that they are alternating copolymers.[9] The copolymerisation proceeds rapidly even though SO_2 does not homopropagate, and neither do the olefins under the reaction conditions, because free radical polymerisation of olefins is unfavourable, and also any alkyl radicals formed will react rapidly with the available SO_2.

The alternating poly(olefin sulphone)s can be classified into four groups depending on the structure of the olefin: (i) 1-olefins, **I**, lead to polymers with a single substituent on the main chain; (ii) 1,1-disubstituted olefins, **II**, such as isobutene, result in a quaternary (disubstituted) carbon atom in the main chain; (iii) 1,2-disubstituted olefins, **III**, such as 2-butene, result in a substituent on each main-chain carbon atom; and (iv) cyclic olefins, **IV**, such as cyclohexene, give ring structures joining adjacent carbon atoms in the chain.

II   III   IV

A variety of substituted olefins, such as allyl alcohol, $CH_2=CH-CH_2-OH$, butadiene, $CH_2=CH-CH=CH_2$,[10] and other dienes, and substituted acetylenes [11] (but not acetylene itself), also form alternating

$$
\begin{array}{ccc}
\text{V} & \text{VI} & \text{VII}
\end{array}
$$

copolymers with SO_2. The poly(acetylene sulphone)s, **V**, have only C=C and C—S bonds in the backbone chain and may be expected to show unusual properties.

2.2. Variable-composition Polysulphones

A number of substituted olefins for which homopropagation of the olefin can compete with addition of SO_2 to the alkyl propagating radical form polysulphones, **VI**, of variable composition, with $\bar{n} \geq 1\cdot0$. In these copolymers, there is a distribution of sequence lengths of the olefin, which varies with the temperature of polymerisation and, to a lesser extent, with the composition of the comonomer feed.

Copolymerisation of ethylene with SO_2 in the gas phase at high pressure gives a polysulphone with $\bar{n} > 1$,[4,12] whereas the copolymer prepared in the liquid phase is alternating.

Vinyl chloride,[13] styrene,[14] choroprene[15] and acrylamide[16] are examples of substituted olefins which form polysulphones of variable composition. The dependence of the structures of these polysulphones on the polymerisation conditions has been examined in detail.

2.3. Copolysulphones or Terpolymers

Two or more olefins can be copolymerised simultaneously with SO_2 to give a polysulphone containing alternating olefin and SO_2 units, with proportions of the different olefins determined by the comonomer feed composition, and particularly the polymerisation temperature.[17-19] There is increasing interest in these terpolymers, as their material properties can be varied, for example,[20] to obtain non-cracking films with high radiation sensitivity.

2.4. Related Polymers

Diallyl monomers undergo cyclopolymerisation with SO_2 to give MS_2 polysulphones containing SO_2 in the ring structure.[21]

Aliphatic polysulphones, **VII**, containing more than two carbon atoms between SO_2 groups in the backbone chain have been prepared by ring-opening polymerisation of cyclic sulphones, or by oxidation of the

VIII

corresponding polysulphides prepared by ionic polymerisation of cyclic sulphides[22] or condensation reactions between dihalides and dimercaptans[23] or addition of dimercaptans to diolefins.[24] They show highly selective scission of C–S bonds during thermal degradation,[25] but do not undergo depropagation.

Polymers containing pendant SO_2 groups, i.e. sulphonyl substituents, could also be considered as polysulphones, although very little has been reported on such polymers,[26] especially their radiation degradation.

The aromatic, or arylene, polysulphones, **VIII**, have become an important class of high-temperature, engineering plastics. They are prepared by step-growth polymerisation and have quite different properties from the poly(olefin sulphone)s. They show high resistance to radiation and predominantly crosslink in vacuum, although the C–S bond is still a site for main-chain scission.[27]

3.   THE CEILING TEMPERATURE AND DEPROPAGATION

3.1. Propagation/Depropagation and Equilibrium

All chemical reactions are, in principle, reversible, and can be treated by equilibrium thermodynamics. Free-radical, chain-growth polymerisations comprise initiation, propagation and termination steps. Propagation consumes monomer to produce polymer and is the step to which the thermodynamic treatment is applied.[28,29]

Polymer will only be formed if the change in the Gibbs free energy function, ΔG, is negative; and at equilibrium ΔG must be zero. ΔG is related to the changes in enthalpy, ΔH, and entropy, ΔS, by eqn (2).

$$\Delta G = \Delta H - T\Delta S = 0 \text{ at equilibrium} \tag{2}$$

During the copolymerisation of olefins with SO_2, there is an increase in order, and hence a decrease in entropy, as the monomer molecules become incorporated into the polymer chain. Consequently, the $-T\Delta S$ term in eqn (2) is positive and ΔH must be negative (exothermic reaction) for polymer to be formed as is indeed the case in the formation of poly(olefin sulphone)s.

Since ΔH and ΔS are both negative, ΔG will become less negative with increasing temperature and eventually a temperature is reached above which ΔG will be positive and long-chain polymer will not be formed. This temperature is known as the ceiling temperature (T_c) and is remarkably low for the copolymerisation of olefins with SO_2, being below 65 °C for most olefins and olefinic compounds, whereas for most vinyl polymerisations it is above 200 °C.

3.2. Concentration Dependence

The ceiling temperature[30] is dependent on the relative proportions of the olefin and SO_2 in bulk polymerisations, and on the monomer concentrations in solution. This occurs through the entropy of mixing of the liquid monomers, according to eqn (3).

$$\Delta S = \Delta S^\circ + R \ln (a(M)a(S)) \tag{3}$$

where $a(M)$ and $a(S)$ are the thermodynamic activities of the two monomers in solution. Since activity coefficients have only been measured for the isobutene/SO_2[31,32] and 3-methyl-1-butene/SO_2[33,34] systems, concentrations are usually substituted for activities.

The relationship between T_c and concentration then becomes

$$T_c = \frac{\Delta H}{\Delta S} = \frac{\Delta H}{\Delta S^\circ + R \ln([M][S])} \tag{4}$$

ΔH can be calculated from the slope and ΔS from the intercept in the plot of $1/T_c$ versus $\ln([M][S])$, since

$$\frac{1}{T_c} = \frac{\Delta S^\circ + R \ln([M][S])}{\Delta H} = \frac{\Delta S^\circ}{\Delta H} + \frac{R \ln([M][S])}{\Delta H} \tag{5}$$

The ceiling temperature typically varies over a 20 °C range in bulk mixtures from 0·1 to 0·9 mole fraction of each monomer, and by much more on dilution of the mixture with solvent.

The ceiling temperature phenomenon was first observed in the copolymerisation of isobutene and SO_2 by Snow and Frey[35] in 1943, but it was not until 1948 that Dainton and Ivin[36] provided the correct thermodynamic explanation.

3.3. Relationship to Olefin Structure

The ceiling temperature, or the equilibrium concentration of monomers, varies greatly with the nature of the olefin. This occurs mainly through the effect of olefin structure on the ΔH term.

Cook *et al.*[37,38] have reported T_c values for a variety of olefins, determined experimentally in SO_2-rich mixtures of olefin and SO_2 and converted to a standard concentration product of $27 \, mol^2 \, litre^{-2}$, corresponding to a mole fraction of olefin of about 0·091. The tabulated T_c values can be converted to equimolar mixtures of olefin and SO_2 using eqn (6) and the experimental values of ΔH, or $-88 \, kJ \, mol^{-1}$ as a general value. ΔS is about $-276 \, J \, K^{-1} \, mol^{-1}$ for all systems.

$$\frac{1}{T_c(2)} = \frac{1}{T_c(1)} + \frac{R}{\Delta H} \ln \left(\frac{[M_2][S_2]}{[M_1][S_1]} \right) \tag{6}$$

The T_c values are higher for 1-olefins than for branched olefins and decrease with increasing size of the olefin (to a limiting value at 1-butene for the 1-olefins). Branching of the alkyl substituent markedly lowers T_c and its effect is greatest when adjacent to the double bond of the olefin. Typical T_c values are given in Table 1.

3.4. Solid Polymer/Gaseous Monomer Equilibrium

The ceiling temperature is of critical importance in determining the thermodynamic stability of poly(olefin sulphone)s towards depropagation. Each poly(olefin sulphone) in the solid state has associated with it a thermodynamically determined equilibrium vapour pressure of olefin and SO_2, which is a function of temperature. This equilibrium will only apply if there are propagating free radicals in the polymer. In the absence of these free radicals, poly(olefin sulphone)s can be thermally stable up to about

TABLE 1
CEILING TEMPERATURES FOR SELECTED
POLY(OLEFIN SULPHONE)S

Olefin	$T_c \, (^\circ C)$
Propylene	90
1-butene	64
Isobutene	5
Cis 2-butene	46
Cyclohexene	24
2-methyl-1-pentene	−34

Data from ref. 37 for monomer mixtures in the liquid state with mole fraction of olefin = 0·091.

200 °C. However, they still have lower thermal stabilities than most vinyl polymers.

The equilibrium vapour pressure of olefin and SO_2 above a poly(olefin sulphone) in the solid state at any temperature can be calculated[5] by extension of the thermodynamic treatment above. Consider the thermodynamic cycle of reactions shown below

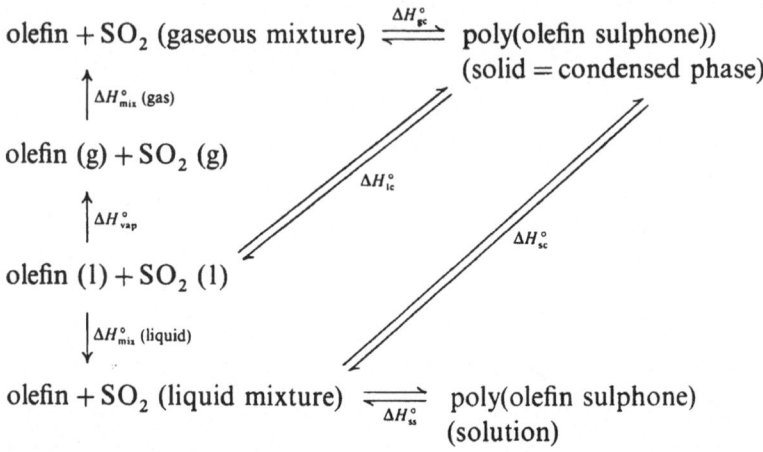

The equilibrium constant K_p for the reaction between a gaseous mixture of olefin and SO_2 and the solid (condensed) polymer is equal to $1/p_m p_s$ where p_m and p_s are the partial vapour pressures of olefin and SO_2, respectively. K_p may be calculated from the standard free energy change (ΔG°_{gc}) via the relationship $\Delta G^{\circ} = -RT \ln K_p$. Thus knowing the value of ΔG°_{gc} enables calculation of the equilibrium vapour pressures of olefin and SO_2. ΔG°_{gc} can be obtained using $\Delta G^{\circ} = \Delta H^{\circ} - T\Delta S^{\circ}$, provided that we can calculate ΔH°_{gc} and ΔS°_{gc}. ΔH°_{gc} may be obtained from the relationship $\Delta H^{\circ}_{gc} = \Delta H^{\circ}_{lc} - \Delta H^{\circ}_{vap} - \Delta H^{\circ}_{mix}$ (of gases), where ΔH°_{vap} is the enthalpy of vaporisation and should be corrected to the required temperature. The relevant thermodynamic data have been reported by Dainton and Ivin.[28] ΔH°_{mix} in the gas phase will be much less than the corresponding term in the liquid phase, and can be taken as $1 \, kJ \, mol^{-1}$. ΔH°_{lc} is obtained by combustion calorimetry of the polymer, or from the relationship $\Delta H^{\circ}_{lc} = \Delta H^{\circ}_{sc} + \Delta H^{\circ}_{mix}$ (of liquids). ΔH°_{mix} is about $6 \, kJ \, mol^{-1}$ in the liquid state.

ΔS°_{gc} can be calculated similarly, or from $\Delta S^{\circ}_{gc} = \Delta S^{\circ}_{ss} - \Delta S^{\circ}_{vap} - \Delta S^{\circ}_{soln}$. The entropy of solution of the polymer in the comonomer mixture will be relatively small and can be neglected.

ΔG°_{gc} is then obtained from the analogue of eqn (2) for standard states,

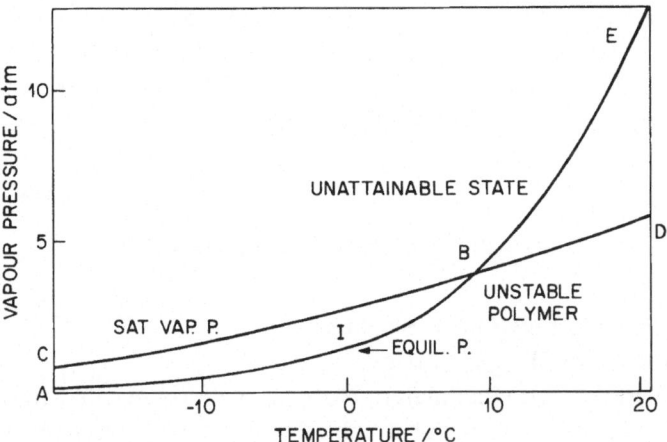

FIG. 1. Calculated temperature dependence of: (i) equilibrium vapour pressure of an equimolar mixture of 1-butene and SO_2 over poly(1-butene sulphone) in the solid state—line ABE; (ii) saturation vapour pressure of an equimolar mixture of 1-butene and SO_2—line CBD.

enabling p_m and p_s to be calculated for any temperature. This calculation is simplified by the fact that, for depropagation of the solid polymer, p_m and p_s can be taken as equal.

These calculations enable the equilibrium vapour pressure of olefin and SO_2 to be calculated as a function of temperature. This is shown for poly(1-butene sulphone) by the curve AE in Fig. 1. Copolymerisation of olefin and SO_2 can occur above this line, whereas depropagation of the polymer can occur below the line. The maximum attainable vapour pressure of the monomers is determined by their saturation vapour pressures and this is shown by the curve CD. These values can be corrected for non-ideality of the liquid mixture of the monomers if the activity coefficients are known. The region above the line CD is unattainable. The crossover point, B, represents the maximum temperature at which the polymer can be thermodynamically stable with respect to depropagation, and the line AB represents the only combinations of temperature and pressure for which the polymer/monomer system is thermodynamically stable. Copolymerisation of olefin and SO_2 in the gas phase can only occur in region I.

It should be noted that the ceiling temperature is lowered by about 60 °C compared to the liquid phase for a 1:1 mixture of the monomers in the gas phase at 1 atm pressure above the solid polymer. Also, it should be recognised that propagating free radicals must be present for equilibrium thermodynamics to apply and that kinetic factors may control the behaviour. Nevertheless, this treatment provides a useful guide to the relative tendency for depropagation of poly(olefin sulphone)s subjected to high-energy radiation as we shall see later.

4. RADIATION DEGRADATION

4.1. Alternating Poly(Olefin Sulphone)s
4.1.1. Molecular Weight Changes
The properties of polymer materials are particularly sensitive to radiation-induced changes in molecular weight. These changes reflect the two main molecular degradation processes: (1) scission of the backbone chains, leading to a decrease in molecular weight; and (2) intermolecular cross-linking of the polymer molecules leading to an increase in molecular weight. As few as one scission or crosslink per molecule can greatly alter solubility, tensile strength and many other properties. The changes in molecular weight can be determined from measurements of solution properties, such as viscosity, osmometry, light scattering and gel permeation chromatography (g.p.c.).

The first observation of γ-radiation causing a decrease in the molecular weight of poly(olefin sulphone)s was probably that of Bray.[39] He prepared poly(olefin sulphone)s from a variety of olefins by copolymerisation with SO_2 using radiation initiation and found that the limiting viscosities of the polymers were lower when higher doses of radiation were used in the preparation, apparently due to the effect of the radiation on the formed polymer. Dainton and Ivin[30] had earlier noted that the viscosity of a solution of poly(1-butene sulphone) was decreased by heating the polymer to 70 °C in solution or in the solid state. Illumination with u.v. light at 25 °C also caused a reduction in viscosity and enhanced the effect of heating at 70 °C indicating that main-chain scission could be induced by u.v. radiation.

In a brief examination of x-ray-initiated grafting of styrene onto poly(1-butene sulphone), Eaton and Ivin[40] found that the viscosity of the polysulphone was reduced, apparently as a result of chain scission by the radiation.

The remarkably high sensitivity of poly(olefin sulphone)s to chain scission by radiation was first discovered and examined quantitatively for poly(1-butene sulphone) and poly(1-hexene sulphone) by Brown and

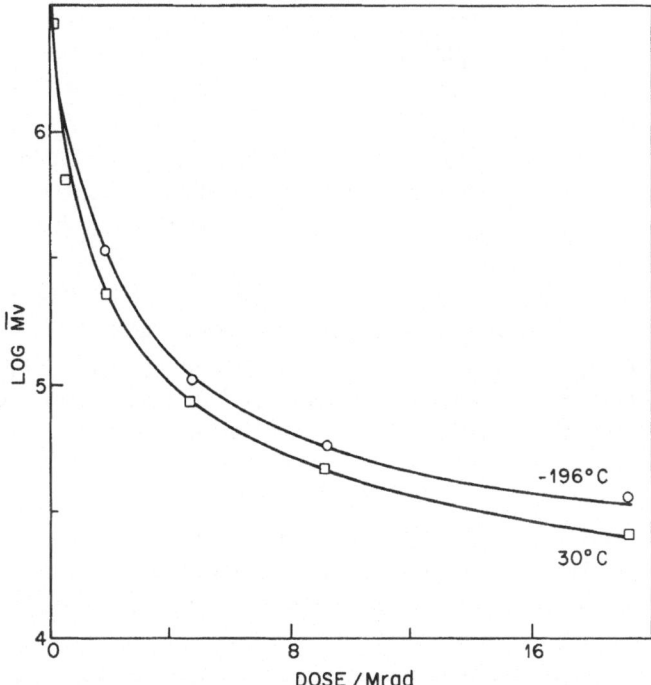

FIG. 2. The reduction in molecular weight ($\bar{M}_v$) of poly(1-butene sulphone) by
γ-irradiation at 30 °C and −196 °C in vacuum.

O'Donnell.[41,42] They measured the molecular weights of these polymers
by viscometry and osmometry after various radiation doses at different
temperatures in air and vacuum. Typical values for the limiting viscosity
number after γ-irradiation of poly(1-butene sulphone) at −196 and 30 °C
are shown in Fig. 2. Similar behaviour was observed for poly(1-hexene
sulphone).

Osmotic pressure measurements are more difficult than viscometry
because (1) the polymer must be soluble in a solvent which is compatible
with the semi-permeable membrane (usually cellulose) of the osmometer,
(2) low molecular weight polymer molecules may permeate the membrane,
and (3) the sensitivity to high molecular weight polymer is low. However,
osmometry has the considerable advantage that it gives absolute values of
$\bar{M}_n$, the number-average molecular weight, and these values can be related
directly to the rate of chain scission or crosslinking according to eqn (7).

$$1/\bar{M}_n(D) = 1/\bar{M}_n(0) + (G(S) - G(X))D/(9 \cdot 65 \times 10^5) \tag{7}$$

Where $M_n(0)$ and $M_n(D)$ are the molecular weights initially and after D Mrad, respectively. A plot of $1/M_n(D)$ versus radiation dose should be linear and the slope will give the value of $G(S) - G(X)$, or $G(S)$ alone if crosslinking is negligible.

Poly(1-butene sulphone) is the first poly(olefin sulphone) in the series of increasing size of the olefin which has a convenient solvent, e.g. acetone, cyclohexanone or tetrahydrofuran, and is therefore a suitable subject for molecular weight studies.

Brown and O'Donnell[42] reported $\bar{M}_n$ values of poly(1-butene sulphone) and poly(1-hexene sulphone) after various doses of radiation in vacuum at 30 °C. From eqn (7), values of $G(S)$ of approximately 10 were obtained for both polymers. A typical plot is shown in Fig. 3. These high G values placed poly(olefin sulphone)s as the polymers with the highest known sensitivity to radiation-induced scission.

In many polymers, scission and crosslinking occur simultaneously. Scission causes the growth of a low molecular tail in the molecular weight distribution (MWD) and crosslinking results in a high molecular tail. Scission alone will lead to a decrease in both $\bar{M}_n$ and $\bar{M}_w$ and an approach

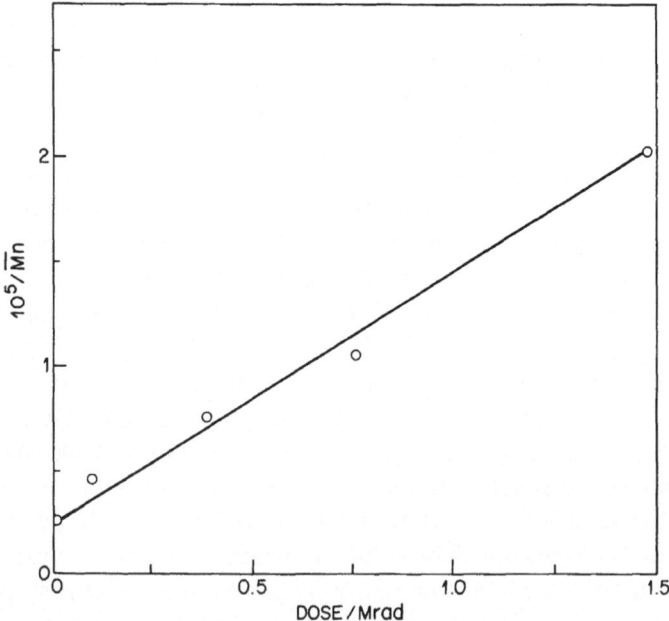

FIG. 3.   Molecular weight $(\bar{M}_n)$ versus dose for γ-irradiation in vacuum of poly(1-butene sulphone) at 30 °C plotted according to eqn (7) in order to determine $G(S)$.

to the most-probable molecular weight distribution with $\bar{M}_w/\bar{M}_n = 2{\cdot}0$, irrespective of whether the initial MWD is narrower or broader than this. Crosslinking will lead to an increase in $\bar{M}_w$ and in the breadth of the MWD. Simultaneous scission and crosslinking can produce either a decrease or an increase in the values of $\bar{M}_w$ and $\bar{M}_n$, and the breadth of the MWD will become greater than $\bar{M}_w/\bar{M}_n = 2{\cdot}0$. Determination of both $\bar{M}_n$ and $\bar{M}_w$ can enable determination of both $G(S)$ and $G(X)$ through the combination of eqn (7), which is applicable to all initial molecular weight distributions, and eqn (8), which is only applicable to the most-probable initial distribution. A modified form of the equation is available for other distributions.[43]

$$1/\bar{M}_w(D) = 1/\bar{M}_w(0) + (G(S) - 4G(X))M_1 D/(1{\cdot}92 \times 10^6) \qquad (8)$$

where M_1 is the molecular weight of a monomer unit.

One method of determining $\bar{M}_w$ is by light scattering. This technique requires considerable experimental expertise; however, the recently developed laser light scattering photometers have simplified experimental complexities thereby greatly improving the potential of this technique for studying molecular weight changes that occur during degradation.

The variation in the entire MWD can be correlated with the values of $G(S)$ and $G(X)$ for any initial MWD.[44] Velocity sedimentation in ultracentrifugation has been used to follow the radiation degradation of some polymers.[45] Gel permeation chromatography has become a very convenient technique for measuring the molecular weight distributions of polymers and hence the values of $\bar{M}_n$ and $\bar{M}_w$. Brown and O'Donnell[42] found for poly(1-butene sulphone) that the breadth of the MWD as well as the individual values of $\bar{M}_n$ and $\bar{M}_w$, as indicated by g.p.c., decreased for initial distributions broader than $\bar{M}_w/\bar{M}_n = 2{\cdot}0$, showing that scission predominated over crosslinking. Some difficulties have been reported in the characterisation of poly(olefin sulphone)s by g.p.c. where they apparently can interact with the column packing.[46]

It is usually considered that crosslinking is negligible in the radiation degradation of poly(olefin sulphone)s of the lower olefins. This assumption does not seem to have been adequately supported in the literature. Gray[47] observed that γ-irradiation of poly(hexadecene-1 sulphone) to a sterilisation dose of $2{\cdot}5$ Mrad produced a substantial increase in density, tensile strength and tensile modulus, indicative of crosslinking and not scission. He suggested that a maximum observed at hexadecene in the variation of various material properties with the length of the alkene might be related to the onset of side-chain crystallisation.[48] O'Donnell[49] has found that side-chain crystallisation does occur in these polymers and is enhanced by orientation.

It is perhaps not unexpected that crosslinking should occur in polysulphones containing large proportions of CH_2 sequences, since the polymer will approximate in part to a polyolefin, viz polyethylene, in which crosslinking occurs readily. It is an interesting, but unresolved, question whether $G(X)$ increases continuously with the size of the 1-olefin, or whether there is a discontinuity at some critical length of the side chain. This could correspond to the onset of side-chain crystallisation. An analogous situation exists in the poly(n-alkyl methacrylate) series, where poly(methyl methacrylate) is considered to undergo negligible crosslinking during irradiation, but n-hexyl and higher alkyl methacrylate polymers develop insolubility, indicating the predominance of crosslinking.

Bowden and Thompson[50] reported a study of the degradation of a variety of poly(olefin sulphone)s irradiated in the form of thin films with 5–20 keV electrons at 20 °C. All samples decreased in thickness, indicating scission and depropagation as for γ-irradiation. They observed a correlation between the ceiling temperatures of the polymers and the rates of degradation, measured by the rates of decrease in film thickness. They deduced that the polysulphones all had similar $G(S)$ values from the doses required for separation of irradiated and unirradiated polymer by differential dissolution.

Bowden[51] has measured the decrease in aliphatic and sulphone absorptions in the infrared spectra of films of poly(2-methylpentene-1 sulphone) with 1 MeV electron beam irradiation. In these experiments SO_2 was lost more rapidly than olefin at lower temperatures, similar to the result of the γ-irradiation studies of Brown and O'Donnell on poly(1-butene sulphone) and poly(1-hexene sulphone). Bowden found that the degradation rate was slower in air than in vacuum and that the difference in rates of loss of SO_2 and olefin was magnified. The rates of degradation were equal above 100 °C and were proportional to the irradiation intensity. A kinetic analysis indicated that the initiation rate was second-order with respect to the sample weight, suggesting initiation of depolymerisation by random chain scission.

4.1.2. Volatile Molecular Products

Irradiation of all polymers produces small radical and ionic fragments, which undergo hydrogen abstraction and combination reactions to form small molecules that are volatile and can be readily separated from the polymer and analysed quantitatively. These volatile products affect the properties of the polymer and also provide direct information on the mechanism of the radiation degradation.

The main volatile products[52] from the irradiation of poly(olefin sulphone)s are the two monomers, sulphur dioxide and olefin. The actual yields and the relative proportions vary with the irradiation temperature as shown in Fig. 4 for poly(1-butene sulphone) and poly(isobutene sulphone). At low temperatures it has been found that SO_2 is the major product, with a G value of about 10, and that the yield of the olefin monomer is quite small. The yields of the SO_2 and olefin increase rapidly with increasing irradiation temperature and the ratio of olefin to SO_2 approaches 1·0. These results are consistent with elimination of SO_2 from the polymer chain at low temperatures and increasing depropagation of both monomers with increasing temperature. The higher depropagation yields from poly(isobutene sulphone) compared with poly(1-butene sulphone) are consistent with the difference in ceiling temperature.

Bowmer and O'Donnell[52,53] found that isomers of the original olefinic

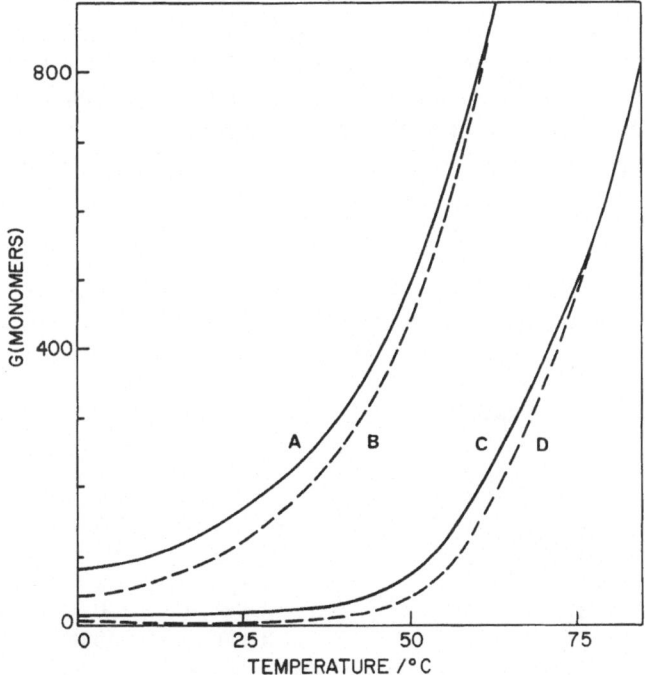

FIG. 4.   Temperature dependence of the yields of olefin (B,D) and SO_2 (A,C) from the γ-irradiation of poly(isobutene sulphone) (A,B) and poly(1-butene sulphone) (C,D) in vacuum.

monomer were produced during radiation degradation. This may occur partly by radiation-induced isomerisation of the initial olefin, but studies with added scavengers[54] do not support this as a major source of the isomers. It was suggested that isomerisation occurred via a cationic chain-end species, formed by scission of the main chain.

Hydrogen is a major product from the radiolysis of most hydrocarbon-containing polymers, but is only a minor product from poly(olefin sulphone)s. A variety of alkane and alkene products were observed, resulting from fragmentation of the main chain and the side chain, followed by combination of the fragments or hydrogen abstraction, but these products were only formed in small amounts. The yields of volatile products from poly(1-butene sulphone) at 0 and 75 °C are shown in Table 2.

4.1.3. Changes in Polymer Structure

The molecular structure, as distinct from the molecular weight of the polymer molecule, may also be changed by irradiation. These changes may include elimination or formation of $C{=}C$ unsaturation and the formation of conjugated sequences, leading to strong chromophores with absorption shifting from the u.v. into the visible region of the spectrum with increasing dose, and elimination or modification of substituents. Such changes are apparently not important in poly(olefin sulphone)s.

One important change does apparently occur in the residual polymer

TABLE 2

YIELDS (G-VALUES) OF VOLATILE MOLECULAR PRODUCTS
FROM γ-IRRADIATION OF POLY(1-BUTENE SULPHONE)[52] AT
TWO DIFFERENT IRRADIATION TEMPERATURES

Product	G values	
	0 °C	75 °C
SO_2	13	475
Butene	3	405
H_2	0·4	0·5
CH_4	0·02	0·02
C_2H_4	0·0002	0·08
C_2H_6	0·01	0·03
Butane	0·02	14
C_5, C_6 hydrocarbons	0·002	0·12
C_8 hydrocarbons	0·15	0·12

after high radiation doses at low temperatures, or an appropriate combination of radiation dose and temperature. More SO_2 is eliminated than olefin causing a change in the composition of the polymer. This may occur either through elimination of SO_2 and recombination of the geminate chain fragments, or by formation of both SO_2 and olefin through depropagation, followed by re-incorporation of the olefin into the polymer. This has been suggested[54] to occur by cationic homopolymerisation of the olefin at a radiation-induced, cationic site on the polymer. This would prevent complete degradation of the poly(olefin sulphone) to its monomers, except under conditions where re-polymerisation could not occur.

4.1.4. Effect of Irradiation Temperature

Temperature is an important variable in radiation degradation, although remarkably few temperature studies have been made on most polymers. Temperature is particularly important in the radiation degradation of poly(olefin sulphone)s, on account of the ceiling temperature phenomenon.

The yields of olefin and SO_2 increase by three orders of magnitude as the irradiation temperature is increased from 0 to 150°C. This is due to increased depropagation and is related to the ceiling temperature for the monomer/polymer system.[55] This is illustrated for nine different poly(olefin sulphone)s in Fig. 5, where the G values for volatile products are plotted on a reduced temperature scale of irradiation temperature/ceiling temperature (in the liquid phase). The T_c values for 1 atm of monomers above the solid polymer will be about 60°C below the liquid phase T_c values and correspond approximately to the temperatures at which $G(SO_2)$ starts to increase from a base level of about 10.

The efficiency of chain scission, measured by $G(S)$, has been reported[42] to show very little dependence on the irradiation temperature. Therefore, it is the depropagation, and not the initiation, which increases with increasing irradiation temperature.

4.1.5. Effect of Irradiation Environment

Fundamental studies of radiation degradation are usually carried out under high vacuum. Irradiation in air, which is a common practical environment, may show greatly enhanced scission and decreased cross-linking, as in polystyrene,[45] or the effect may not be great, as in poly(methyl methacrylate). The effect of oxygen is evidently related to the chemical structure of the polymer and its physical form, e.g. thick sheet or fine powder. Limited results on poly(olefin sulphone)s suggest that air causes a small increase in scission,[42] but depropagation may be reduced.[51]

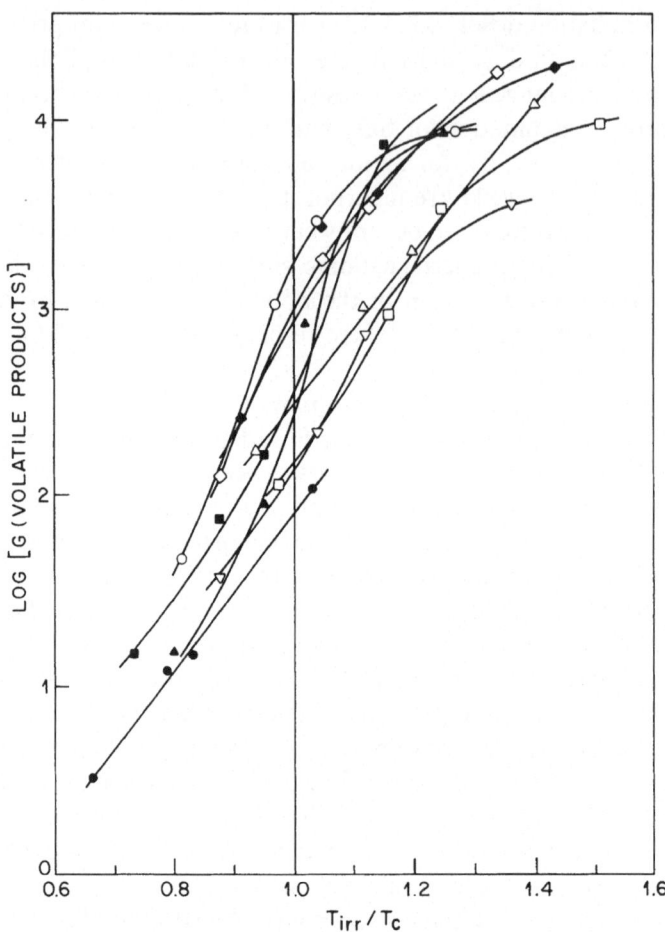

FIG. 5. Temperature dependence of the initial rates of formation of volatile products during the γ-irradiation of various poly(olefin sulphone)s in the solid state. T_{irr} is the irradiation temperature. (●) Poly(ethylene sulphone); (■) poly-(propylene sulphone); (▲) poly(1-butene sulphone); (□) poly(isobutene sulphone); (○) poly(1-hexene sulphone); (△) poly(4,4-dimethyl-1-pentene sulphone); (▽) poly(3-methyl-1-butene sulphone), (◆) poly(2-butene sulphone); (◇) poly(cyclo-hexene sulphone). (Taken from ref. 55.)

4.2. Variable Composition Polysulphones

The radiation degradation of poly(olefin sulphone)s with $\bar{n} > 1$ has only been reported for poly(styrene sulphone)[56] and some preliminary work on poly(vinyl chloride sulphone)[57] has yet to be published.[58]

As the proportion of olefin sequences with $\bar{n} > 1$ increases, and their average length increases, the amount of depropagation must decrease and the radiation degradation should show features of the corresponding poly(olefin). Although C–S scission should always be a favoured reaction on account of the low strength of this bond, $G(SO_2)$ may decrease in the absence of a repulsive effect from a penultimate SO_2 group, and depropagation will be inhibited.

Stillwagon et al.[56] reported a $G(S)$ of 1·1 for poly(styrene sulphone) with $\bar{n} = 2$. This is considerably lower than the $G(S)$ of 10 reported for the poly(olefin sulphones) and presumably reflects the protective effect of the benzene ring. The $G(S)$ values of the higher styrene content polymers ($\bar{n} > 2$) were not reported (polymers with $\bar{n} < 2$ have not been studied, possibly because of solubility limitations in common solvents; the alternating copolymer ($\bar{n} = 1$) appears to be insoluble). Since poly(styrene) crosslinks in vacuum under γ-irradiation, one might expect that $G(S)$ would decrease with increasing styrene sequence length and $G(X)$ would increase. The trend in lithographic sensitivity which decreases as $\bar{n}$ increases supports this hypothesis.

A preliminary study[58] of the degradation of poly(vinyl chloride sulphone) by γ-radiation has shown that SO_2 elimination is the main reaction at low temperatures, but that there is considerable dehydrochlorination at higher temperatures. Development of unsaturation, which would be expected from dehydrochlorination, was shown by an increasing $C=C$ absorption at $1610\,cm^{-1}$ in the infrared spectrum. Some acetylene and ethylene were also observed in the radiolysis products, indicating depropagation of dehydrochlorinated units, or secondary radiolysis reactions of vinyl chloride produced by depropagation. The facile dehydrochlorination of single vinyl chloride sequences between SO_2 groups has been demonstrated by Cais and O'Donnell.[57]

A surprising feature of the radiolysis was the apparent absence of vinyl chloride. However, Cais[59] has shown that there are no single vinyl chloride sequences in this polysulphone with $\bar{n} \geq 2$, and the polymer would therefore not be expected to undergo depropagation to vinyl chloride. Dehydrochlorination of single vinyl chloride sequences occurs very rapidly during polymerisation with radiation initiation, and during storage at room temperature in the solid state and in solution in some solvents.

5. MATERIAL PROPERTIES

Studies of the copolymerisation of olefins with sulphur dioxide have made a significant contribution to the fundamental understanding of polymerisation kinetics and thermodynamics. Determining the origin of the ceiling temperature effect and its relation to molecular structure has been especially important. But poly(olefin sulphone)s also exhibit a diversity of physical, chemical and mechanical properties which make them potentially suitable for a range of applications.[3] The inexpensive raw materials and the ease of copolymerisation have provided a strong commercial incentive, but despite a vast amount of research, no commercial development of these polymers has been forthcoming. This contrasts with the aromatic poly-sulphones, which have been successfully exploited on a commercial scale and constitute an extremely important class of engineering thermoplastics.[60]

The primary reason for the lack of success in developing poly(olefin sulphone)s as commercial thermoplastics has been their poor thermal stability. These polymers are generally characterised by decomposition temperatures that are below the temperatures necessary for fabrication by conventional moulding techniques. For example, the softening temperature of poly(1-butene sulphone) is reported to be 160 °C,[61] but the polymer begins to decompose thermally at temperatures as low as 130 °C.[30,62] Consequently, degradation will occur at the temperatures required to mould the polymer.

Such low decomposition temperatures are not necessarily implied from the low ceiling temperatures associated with formation of these polymers although the same kinetic and thermodynamic considerations associated with the ceiling temperature phenomenon may strongly influence the decomposition rate once the reaction has been initiated. The ceiling temperature pertains to the equilibrium between monomer and propagating species. Although a polymer may be thermodynamically unstable with respect to its monomer above T_c, it will not depolymerise spontaneously but will only do so under conditions that produce propagating species. The formation of propagating radicals by chain scission of the polymer at elevated temperatures provides a kinetic pathway for depropagation, and would be expected to lead to considerable evolution of monomer in an attempt to establish equilibrium, with consequent high rates of degradation.

The radiation degradation studies of O'Donnell and his co-workers[41,42,52,55,63] discussed in the first part of this review further demonstrate the inherent instability of these polymers. The carbon–sulphur bond in the main chain is relatively weak (bond energy $\simeq 60\,\mathrm{kcal\,mol^{-1}}$)[1] and undergoes selective chain cleavage upon γ-irradiation with a $G(S)$ of 7–12.[41,56] This value of $G(S)$ is considerably greater than that of most other chain-degrading polymers and is in contrast to the radiation stability shown by the aromatic polysulphones, which crosslink with an extremely low G value.[27]

6. THE LITHOGRAPHIC PROCESS

The degradation of poly(olefin sulphone)s under irradiation proved to be advantageous for lithographic pattern delineation as practised by the microelectronic circuit industry in the manufacture of integrated circuits. These circuits are made by diffusing small amounts of impurities, such as boron or arsenic, into specific regions of a semiconductor substrate, e.g. silicon, to produce the desired electrical characteristics of the circuit. The doped regions (together with the conductor paths that link the active circuit elements) are defined by lithographic processes[64] in which the desired pattern is first generated in a resist layer (usually a polymeric film $\simeq 0 \cdot 5\text{--}1 \cdot 0 \, \mu m$ thick which is spin-coated onto the substrate) and then transferred via processes such as etching to the underlying substrate. In silicon integrated circuit manufacture, the substrate is usually silicon dioxide, which is present as a thin layer on top of the silicon and functions as the actual mask for the subsequent diffusion process.

The definition of the pattern in the resist layer is achieved by selectively exposing the resist to some suitable form of radiation such as u.v. light, electrons, x-rays or ions. The resist contains radiation-sensitive groups which chemically respond to the incident radiation forming a latent image of the circuit pattern, which can subsequently be 'developed', e.g. by solvent treatment, to produce a three-dimensional relief image in the resist. Poly(olefin sulphone)s would appear to be well-suited to this application since they degrade by chain scission and the lower molecular weight of the degraded polymer molecules should enable the exposed regions to be differentiated from the unexposed regions on the basis of differences in dissolution rate. Such materials are called positive resists as opposed to negative resists which crosslink on irradiation and give rise to a negative image in the resist after development. The areas of resist remaining after development must now protect the underlying substrate during the variety of additive and/or subtractive processes encountered in semiconductor processing. If, for example, the underlying substrate were silicon dioxide, immersion of the structure into an etchant such as buffered hydrofluoric acid would result in selective etching of the SiO_2 in those areas that were bared during the development step.

The basic steps in the process are shown schematically in Fig. 6. The example shown corresponds to photolithography in which the photo-sensitive resist, or photoresist as it is called, is applied as a thin film to the substrate and subsequently exposed in an image-wise fashion through a mask. The mask contains clear and opaque features that define the circuit

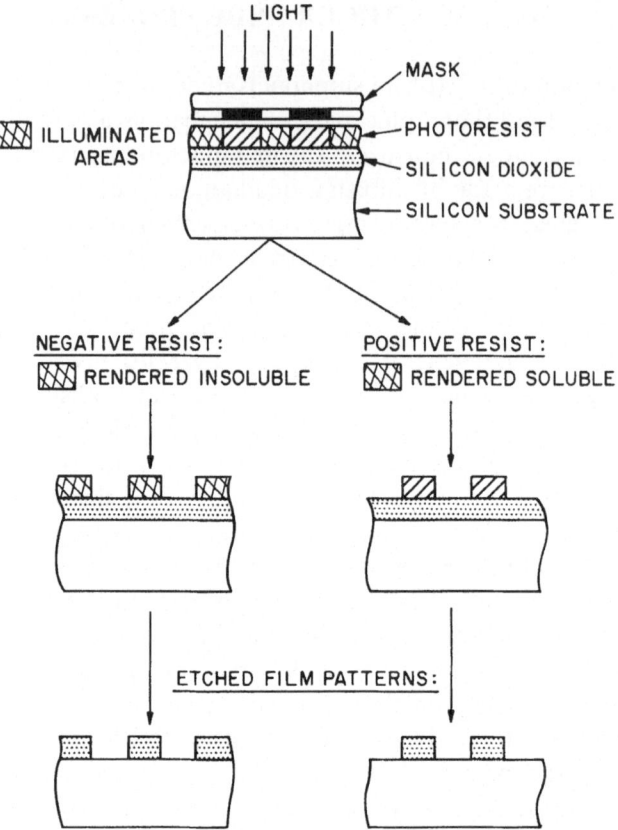

FIG. 6. Schematic diagram of the photolithographic process showing exposure and development of negative and positive resists and the resulting etched film patterns.

pattern. Alternatively, the resist could be exposed to x-rays (x-ray lithography), electrons (electron-beam lithography) or even ions (ion-beam lithography). These latter technologies are becoming important as photolithography approaches resolution limits set by diffraction. Since x-ray wavelengths are typically in the 4–10 Å range (the deBroglie wavelength for electrons is considerably smaller still), diffraction effects are not important and resolution is instead limited by other effects such as scattering. Focused electron- and ion-beam lithography also have an advantage in being maskless technologies. The beam is focused to a spot and scanned across the substrate while being modulated, i.e. turned on and off (blanked), under control of a computer. It may be noted however that, irrespective of the exposure technology, a resist is required as the primary imaging medium.

The role of the resist is seen to be twofold. First, it must respond to the exposing radiation forming a latent image of the circuit pattern which can subsequently be developed to produce a three-dimensional relief image. Second, the areas of the resist remaining after development (the exposed or unexposed areas as the case may be) must protect the underlying substrate during subsequent processing.

7. ELECTRON BEAM RESISTS

The development of electron-beam lithography in the early 1970s necessitated a parallel development of suitable resists with sensitivities commensurate with machine design parameters. This was particularly true of positive resists where most of the early research had concentrated on poly(methyl methacrylate), which had a sensitivity at least an order of magnitude lower than that required. Poly(olefin sulphone)s on the other hand appeared to possess many of the properties required of a positive electron resist. O'Donnell and his coworkers had already demonstrated that these polymers degraded under irradiation by chain scission as required for development by fractional dissolution, with $G(S)$ values significantly greater than most other chain-degrading polymers. The response to electron irradiation had not been determined, but studies on other polymers had indicated comparable sensitivity to electron and γ-radiation indicating that the polysulphones should also be readily degraded by electron irradiation. They were known to be amorphous polymers, soluble in a wide variety of organic solvents and capable of being deposited as uniform films on a substrate by spin-coating. They were also known to be resistant to many of the chemical etching solutions encountered in semiconductor processing.

The lithographic properties of poly(olefin sulphone)s were first reported by Bowden and Thompson at Bell Laboratories.[50,65] They investigated polysulphones based on a variety of aliphatic and cyclic olefins and found that the high $G(S)$ values previously determined from γ-radiolysis studies were indeed reflected in their electron sensitivities, which were of the order of $1\cdot0\,\mu\mathrm{C\,cm^{-2}}$ at $10\,\mathrm{kV}$. This was a factor of 10 better than resists based on poly(methyl methacrylate) and was commensurate with the sensitivity required by electron beam writing machines under development at that time. Consideration of all the material properties required of a resist led

Bowden and Thompson to select poly(1-butene sulphone), PBS, as the most promising resist candidate.[66,67] This polymer produced uniform, pinhole-free films when spun from various solvents and showed good etch resistance to both acidic (buffered hydrofluoric acid) and basic (KOH/Fe_3CN_6) etching solutions. Also, its glass transition temperature (T_g) was close to 100 °C (a high T_g is desirable from the standpoint of minimising distortion during solvent development).[67]

The lack of thermal stability, which is characteristic of all poly(olefin sulphone)s, precluded the use of dry-etching techniques such as plasma- or reactive-ion etching, rendering PBS unsuitable for direct-write fabrication of integrated circuits where dry-etch resistance is required in order to take advantage of the high-resolution capability afforded by electron-beam lithography. Nevertheless, PBS did appear to be ideal for electron-beam fabrication of the chromium photomasks used in photolithography. Since the chromium film is only a few hundred angstroms thick, wet etching is perfectly adequate and undercutting which is encountered with thick ($> 0.2 \mu m$) films is not a problem.

7.1. Factors Affecting Resist Performance

An extensive programme was undertaken at Bell Laboratories to develop PBS as a positive resist for making chromium master masks by electron-beam lithography using the EBES electron beam exposure system also developed at Bell Laboratories. This program involved extensive investigation in the areas of synthesis, solution properties, thermal properties and processing characteristics. Most of the details have, however, remained proprietary, although some general comments may be made which are pertinent to optimising the performance of all positive resists.[68]

7.1.1. Molecular Weight

The molecular weight of the polymer needs to be as high as practically allowable. It can be shown that the fraction (p) of bonds broken by exposure of a resist film of thickness (z) to an incident electron dose (D) in $C\,cm^{-2}$ is given by[69]

$$p = \frac{EG(S)DM_1}{100qN\rho z} \tag{9}$$

where q is the electronic charge, ρ is the density of the polymer, E is the energy absorbed in the film per incident electron, M_1 is the molecular weight of a monomer unit and N is Avogadro's number. The fraction of bonds broken should be sufficient to reduce the molecular weight to a value such that there is almost complete separation of the molecular weight distributions of the irradiated and unirradiated polymer. This facilitates an 'ideal' development condition, wherein there is negligible thinning of the unirradiated resist during solvent development.

The molecular weight of the irradiated polymer is related to that of the unirradiated polymer by eqn (7). By combining eqns (7) and (9), the molecular weight of the irradiated polymer can be expressed by eqn (10).

$$\bar{M}_n(D) = \frac{\bar{M}_n(0)}{1 + \left(\dfrac{EG(S)}{100qN\rho z}\right)D\bar{M}_n(0)}$$

$$= \frac{\bar{M}_n(0)}{1 + KD\bar{M}_n(0)} \tag{10}$$

where K is a constant for a fixed beam accelerating voltage and film thickness. Figure 7 shows a plot of $\bar{M}_n(D)/\bar{M}_n(0)$ versus $\bar{M}_n(0)$ for various radiation doses. If we arbitrarily assume that the development condition corresponds to the scission of one tenth of the bonds in a polymer chain (i.e. an order of magnitude decrease in molecular weight), then, as seen in Fig. 7, the sensitivity will increase with increasing molecular weight. As the molecular weight increases, so does the solution viscosity, and this will

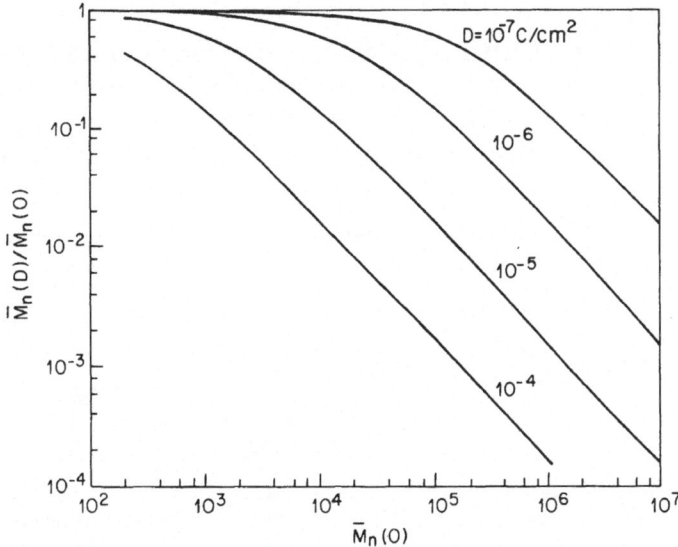

FIG. 7.   Fractional change in molecular weight as a function of initial molecular weight for different incident doses.

eventually give rise to problems in solution filtration and spin coating, so that there must be a compromise between sensitivity and processability. In general, molecular weights of the order of 10^5–10^6 appear practical.

7.1.2. Molecular Weight Distribution (MWD)

For random scission, the MWD of the degraded material should approach $\bar{M}_w/\bar{M}_n = 2$ and be independent of the initial distribution.[70] Figure 8 shows the distributions obtained after 3 Mrad of γ-irradiation for two samples of PBS with quite different initial distribution. Both samples give the same distribution after 3 Mrad in accordance with theory, but the irradiated and unirradiated distributions are not yet separated in the case of the sample with the broad initial distribution. As a general rule, the broader the initial distribution, the greater will be the overlap of the MWD of the irradiated and unirradiated polymer for the same radiation dose. The copolymerisation of olefins and SO_2 is a free radical, chain-growth reaction and therefore the minimum MWD of the polymerised material will be 1·5–2·0. Although it might be desirable to reduce this by fractionation, such an approach is not really economically practical.

PBS with a molecular weight of the order of 10^6 and a MWD of less than 2·5 only requires 3–4 Mrads to produce sufficient shift in the distribution to enable solvent development. This absorbed dose translates into an incident electron dose of 0·8 $\mu C\,cm^{-2}$ at 10 kV which is commensurate with EBES requirements.[67]

Since the first reports by Bowden and Thompson on the application of poly(olefin sulphone)s as electron resists, several papers have been published by groups at IBM[20,71,72] and RCA[73,74] on various lithographic aspects of these materials. But in spite of the large amount of work devoted

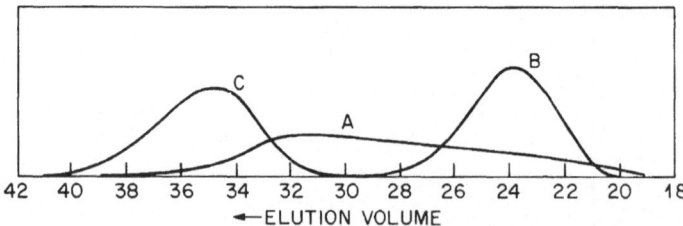

FIG. 8.   Gel permeation chromatography traces showing the effect of initial molecular weight distribution (MWD = M_w/M_n) on separation dose: (A) whole polymer prepared by u.v. initiation (MWD = 24·4); (B) polymer obtained by fractionation of (A) (MWD = 1·7); (C) polymer obtained from both (A) and (B) after 3 Mrad (MWD = 2·2).

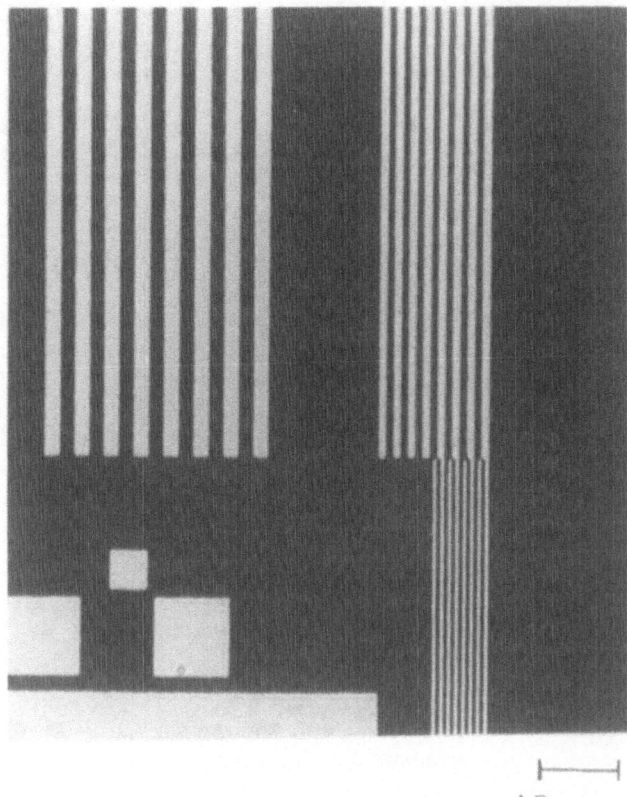

FIG. 9.   Transmission photomicrograph showing $2\cdot0\,\mu m$, $1\cdot0\,\mu m$ and $0\cdot5\,\mu m$ line and space structures etched in chromium.

to developing new materials, PBS has remained unchallenged as the resist of choice for mask making. This is probably more due to economic considerations and to the fact that PBS was introduced first than to any unique property. The resist was introduced commercially in 1976 and at present about 75 % of masks produced in the USA by electron lithography are fabricated using PBS resist.

7.2. Resolution

The resolution capabilities of PBS are demonstrated in Fig. 9. Minimum features of $1\cdot0\,\mu m$ can be achieved routinely without exposure or developing compensation. With care, features of the order of $0\cdot5\,\mu m$ can be written and even $0\cdot25\,\mu m$ resolution has been achieved. Figure 10

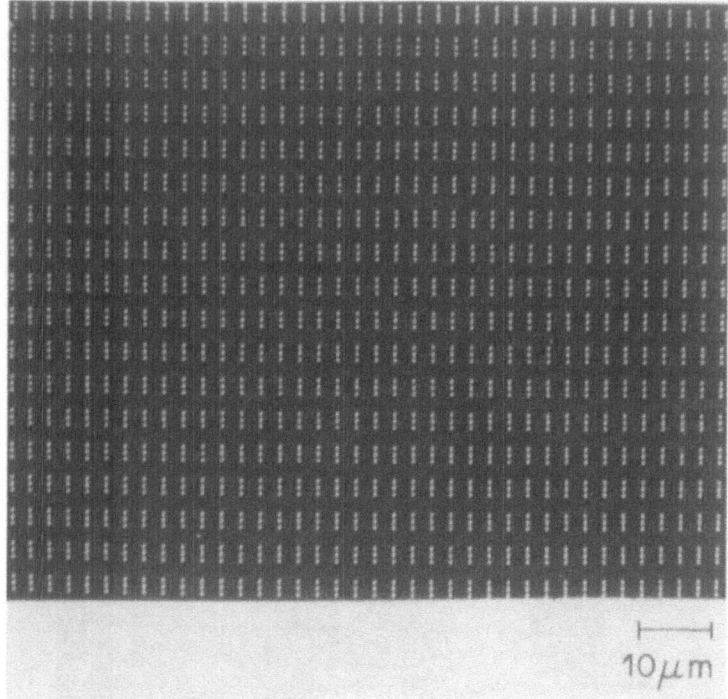

FIG. 10. Transmission photomicrograph showing an array of three $0.5 \mu m$
windows on $1.0 \mu m$ centres.

shows an array of dots nominally $0.5 \mu m$ in diameter on a mask that was
used to fabricate Schottky barrier diodes as part of a programme to extend
the frequency range of cryogenically-cooled millimetre-wave receivers to
230 GHz.

The resolution capabilities of PBS also allow evaluation of the lithographic writing tool. Figures 11 and 12 are SEM micrographs that show
various EBES characteristics. All of the micrographs show etched chromium
features, the PBS having been stripped. The $0.5 \mu m$ address structure of
EBS is clearly visible on feature edges inclined at $45°$ to the scan direction
(Fig. 11). The lines in Fig. 12(a) are aligned parallel to the sweep direction
of the beam while the lines in Fig. 12(b) are aligned perpendicular to the
sweep direction. The rougher edges in the case of the horizontal lines can
be explained by the fact that the beam blanking pattern results in a
significantly lower exposure dose for single address horizontal lines.
Reduced exposure gives rise to the periodic edge ripple displayed in Fig.
11(a). With overdevelopment, these edges can be made smoother.

The quality, or edge acuity, of developed resist features depends upon

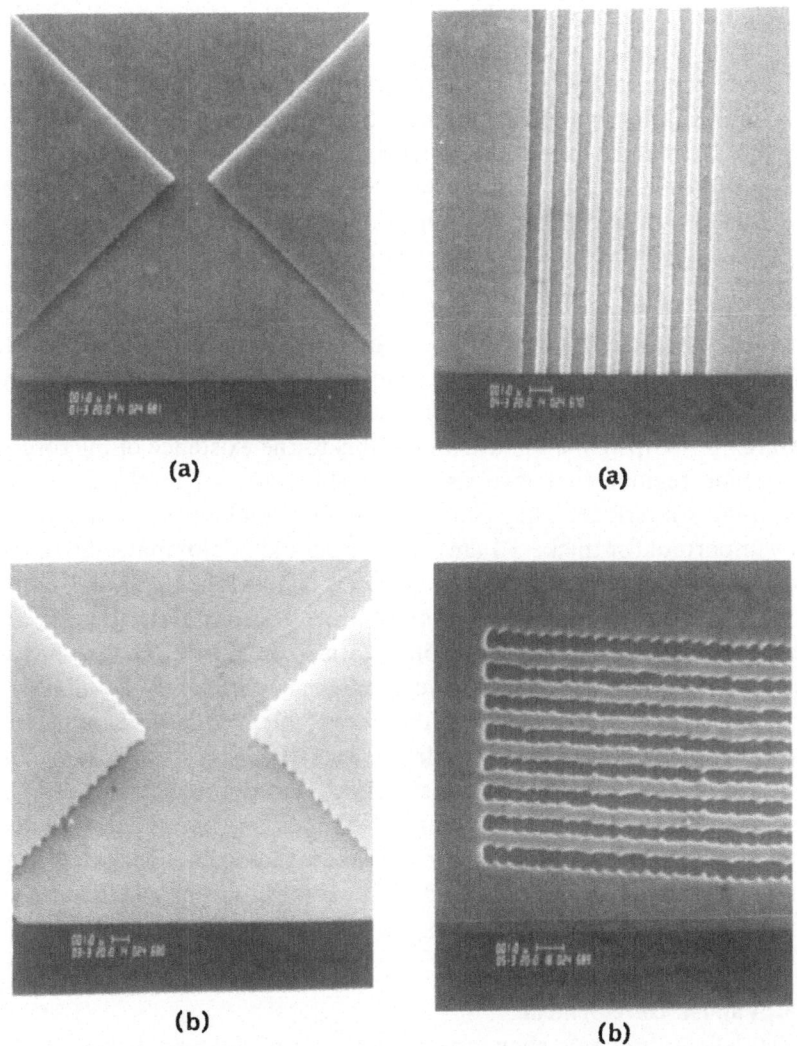

(a)

(a)

(b)

(b)

FIG. 11. SEM micrographs showing 0·5 μm address structure detail on etched chromium features inclined at 45° to the scan direction.

FIG. 12. SEM micrographs of etched 0·5 μm lines which are (a) vertical (lines written parallel to the beam sweep direction) and (b) horizontal (lines written perpendicular to the beam sweep direction).

the extent of distortion introduced during the development process. Development by fractional dissolution invariably involves swelling of the polymer matrix, although the degree of swelling can be minimised by judicious choice of solvent and temperature. The developing solvent may be a single solvent or a solvent mixture. Such mixtures generally consist of a 'good' solvent and a 'poor' solvent with the proportions adjusted to give an acceptable difference in dissolution rate between exposed and unexposed areas. However, most information in this area is proprietary.

7.3. Film Cracking

One of the contentious points concerning application of poly(olefin sulphone)s to lithography is their reported tendency to crack during film formation and particularly during solvent development following exposure.[71,72] Gipstein et al. attributed this to the existence of microscopic crystalline regions that give rise to swelling stresses at the crystalline/ amorphous interface. They claimed that this cracking phenomenon was only important for thick ($>1 \mu m$) films and could be eliminated by suitable choice of a comonomer. For example, poly(cyclopentene-co-1-butene-sulphone) was reported to give crack-free, $1 \cdot 25 \mu m$ resist images after development at an exposure dose of $4 \mu C \, cm^{-2}$ at $25 \, kV$. Cracking has also been reported in poly(1-methyl-cyclopentene sulphone) by Poliniak et al. at RCA,[75] although they claimed that the problem could be eliminated by carefully filtering the resist and drying the spun film in a high vacuum, followed by storage in a moisture-free environment. Bowden and Thompson did not observe cracking in PBS, possibly because they confined their attention to film thicknesses of the order of $0 \cdot 5 \mu m$, which is all that is required for mask making. They did, however, report cracking in $0 \cdot 5 \mu m$-thick films of polysulphones prepared from styrene/SO_2 mixtures.[76]

7.4. Vapour Development

Poly(olefin sulphone)s show another interesting feature during electron beam irradiation. Bowden and Thompson[77,78] found that the film thickness decreased on irradiation at a rate determined primarily by olefin structure, dose rate (beam current) and temperature. Other parameters, such as molecular weight and accelerating voltage, were of minor importance.[79] They termed this phenomenon 'vapour development' since, under appropriate conditions, the resist could be totally removed, thereby removing the need to dissolve the irradiated region, which is an attractive feature for certain applications. It was found that the rate of vapour

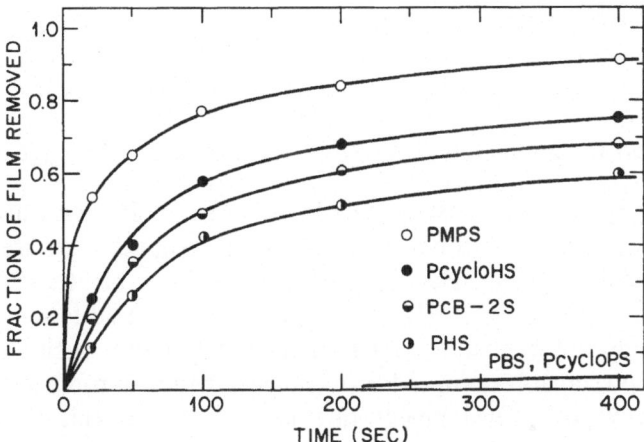

FIG. 13.   Effect of olefin structure on the rate of vapour development of various poly(olefin sulphone)s. PMPS—poly(2-methyl-1-pentene sulphone); PcycloHS—poly(cyclohexene sulphone); PcB-2S—poly(cis-2-butene sulphone); PHS—poly-(hexene-1-sulphone); PBS—poly(butene-1 sulphone); PcycloPS—poly(cyclopentene sulphone).

development was markedly dependent on the olefin structure. As seen in Fig. 13, the initial rate increases as the ceiling temperature for polymerisation decreases. It will be noted that PBS has a negligible rate of vapour development at room temperature, although at elevated temperature the rate approaches that of poly(2-methyl-1-pentene sulphone), PMPS (see Fig. 14).

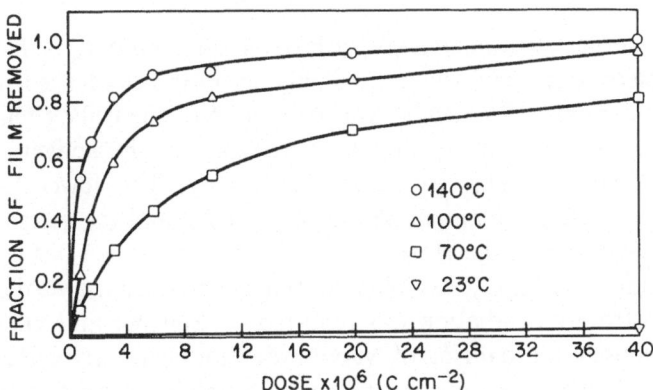

FIG. 14.   Effect of irradiation temperature on the rate of vapour development of PBS (current $= 5 \times 10^{-10}$ A, voltage $= 5\,$kV).

Vapour development is a result of the ceiling temperature phenomenon discussed earlier. The volatile radiolysis products consist mainly of the parent monomers and the large G values result from chain depropagation[80]

$$\text{polysulphone (condensed)} \xrightarrow{k_d} \text{olefin (gas)} + SO_2 \text{ (gas)} \qquad (11)$$

where k_d is the rate constant for depropagation. Radicals are formed continually during electron irradiation and the product gases are removed by the vacuum system. Hence, equilibrium will not be attained and the reaction should move entirely to the right, assuming that terminating side reactions do not interrupt the depropagation reaction. Such terminating reactions presumably occur in PBS during irradiation at room temperature, since the vapour development rate is very low at this temperature. Accordingly, one might expect that the rate of vapour development should depend on the rate of depropagation, since the faster a chain unzips the less likely it is that secondary reactions such as termination or transfer will terminate the depropagating chain. This apparently is the case in PBS for which the vapour development rate increases rapidly with increasing temperature. Further, assuming that the rate constant can be expressed in terms of an Arrhenius expression, $k_d = A \exp(E_d/RT)$, where E_d is the activation energy for depropagation, then the different vapour development rates for the various poly(olefin sulphone)s should reflect the effect of olefin structure on k_d.

The activation energy for depropagation is related to the enthalpy change for the reaction via the expression

$$E_d = E_p - \Delta H_{gc} \qquad (12)$$

where E_p is the activation energy for propagation and ΔH_{gc} is the enthalpy of polymerisation for formation of polymer in the condensed state from gaseous monomer. Since ΔH_{gc} can be related to the ceiling temperature (eqn (4)), variations in T_c should be reflected in the activation energy for depropagation (E_d) to gaseous monomer. The lower the ceiling temperature, the smaller will be the value of E_d, and the rates of vapour development should therefore increase with decreasing T_c as is observed experimentally (Fig. 13). A similar correlation between the yield of volatile products from γ-irradiation was reported by Bowmer and O'Donnell.[52] The proposed mechanism of vapour development, involving random scission followed by depropagation with termination by recombination, was further substantiated from kinetic studies.[78] The overall activation energy for PBS was 41 kJ mol^{-1} and 28 kJ mol^{-1} for PMPS.

7.5. Dry Etching

The major limitation of poly(olefin sulphone)s in resist applications is their poor resistance to dry etching. Clearly, a fast, vapour-developing resist such as PMPS would have no resistance to dry-etching environments, since the active species present in the plasma have sufficient energy to initiate chain scission with subsequent rapid material loss. But even PBS, which is characterised by a negligible rate of vapour development at room temperature, exhibits poor dry-etching resistance. This results from the elevated temperatures (where the vapour development rate of PBS is high) attained in a plasma environment and suggests that plasma-etch resistance

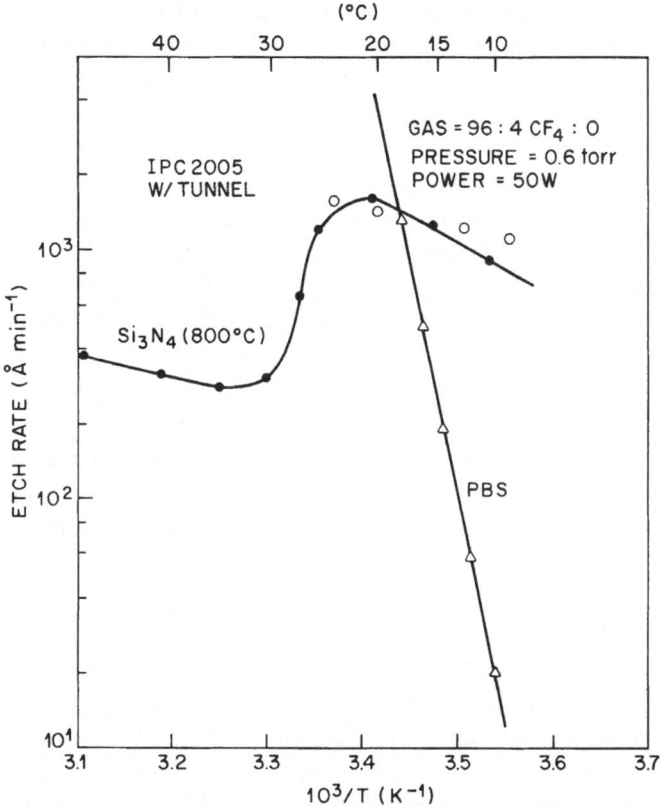

FIG. 15.   Temperature dependence of plasma etch rates of PBS ($\triangle$) and silicon nitride ($\bullet$). The etch rates of silicon are shown as open circles ($\bigcirc$). Plasma etching was carried out in an IPC barrel reactor model 2005 equipped with a perforated aluminium shield (tunnel) to minimise the effects of ion bombardment. A 96:4 gas mixture of CF_4 and O_2 at a pressure of 0·6 torr and a power of 50 W was used as the etchant.

could be improved by reducing the temperature of the substrate. This approach has been partially successful.[81] As seen in Fig. 15, the plasma-etch rate of PBS is lower than either silicon or silicon nitride below about 20 °C, and by cooling the substrate and etching below this temperature, it is possible to selectively etch such substrates.

Another approach to improving resistance to dry-etching environments involves chemically altering the resist prior to etching. Himics *et al.*[82] recently reported that exposure of poly(olefin sulphone)s, particularly those derived from cyclopentene, to u.v. radiation enables the resist to withstand subsequent sputter etching. Yamazaki *et al.*[83] have also reported a reversal gas-etching technique for chromium films using PBS. The technique depends on the presence of a layer of tungsten oxide in the surface of an antireflective chromium film. The latter consists of a layer of chromium oxide on a glass substrate. The WO_3 layer acts as a masking layer in the gas plasma environment (mixture of CCl_4, N_2 and O_2). Thus the areas exposed following exposure and development of PBS are masked during plasma treatment, whereas in the unexposed areas the residual resist is rapidly removed. Ordinarily, plasma etching would halt at the interface, i.e. at the WO_3 layer. However, the WO_3 is apparently removed in this region by interaction with the decomposition products of the resist and thus etching continues through the chromium film resulting in a negative-tone mask pattern.

These techniques are extremely difficult to control in practice and have found little utility. Lowering the substrate temperature is practised in those instances where only a mild plasma treatment is required, such as descumming (this process is designed to remove residual traces of resist from exposed and developed areas), but it is not satisfactory for etching thick substrate films.

7.6. Composite Systems

A novel approach to circumventing this problem of poor etch resistance in the application of poly(olefin sulphone)s to lithography was reported by Bowden and coworkers.[84,85] They proposed a two-component resist system consisting of a novolac matrix resin similar to that used in common quinonediazide positive photoresists, and a poly(olefin sulphone) dissolution inhibitor, or sensitiser, in solid solution in the novolac. The novolac resin is soluble in aqueous, basic solutions, but its rate of dissolution in the base is markedly reduced when the dissolution inhibitor is present.

Bowden *et al.* chose PMPS as the dissolution inhibitor, the idea being that it would depolymerise on irradiation and thus be removed from the

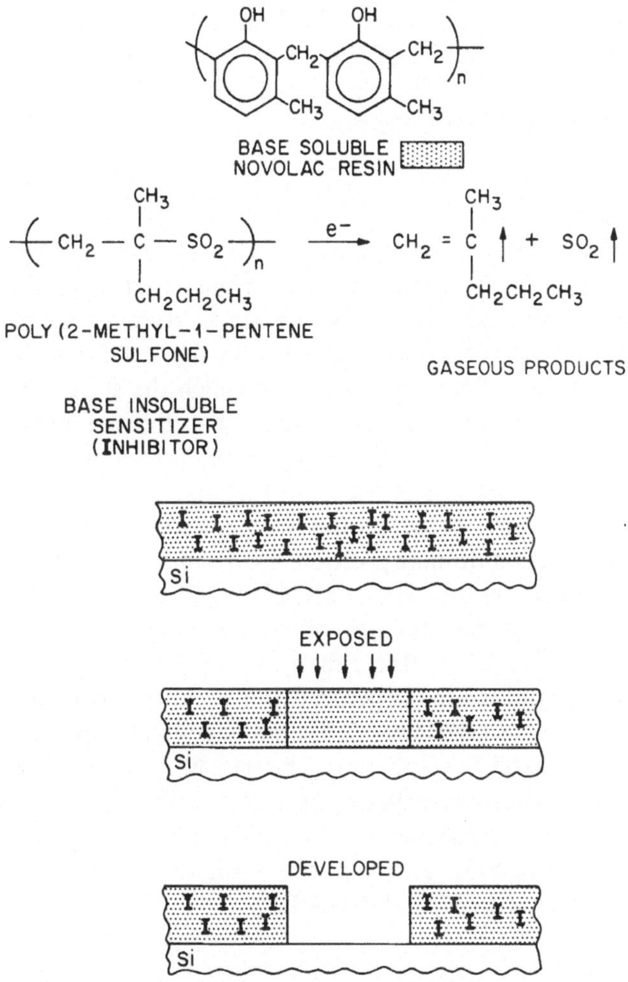

FIG. 16. Schematic representation of NPR (novolac-based positive electron resist) process.

matrix with a concomitant increase in solubility of the remaining film, which now consists largely of pure novolac resin (see Fig. 16). The sensitivity of this resist, which is called NPR, is $3–5 \, \mu C \, cm^{-2}$ at $20 \, kV$. The resist succeeds in combining the favourable characteristics of both components, viz the excellent dry-etch resistance of novolac polymers and the high sensitivity of the polysulphone sensitiser.

This system has generated considerable interest and developments along the same conceptual design have been reported subsequently by Shiraishi *et al.*[86] at Hitachi Chemical and by Willson and coworkers at IBM.[87] The latter modified the structure of PMPS by incorporating 20–80 mol % of an alkyl-1-alkoxyalkylethylene, such as 2-ethoxyethylmethallyl ether, to improve compatibility with the novolac.

7.7. Aromatic Groups

An alternative approach to building dry-etch resistance into the poly(olefin sulphone) molecule is to incorporate aromatic groups. Bowden and Thompson[76] prepared polysulphones from styrene, which may be synthesised with compositions ranging from the 1:1 copolymer through to homopolystyrene. Only copolymers in which the styrene/SO$_2$ ratio is greater than 2 are soluble. They found that the 2:1 copolymer was thermally stable up to 250 °C and exhibited high resistance to dry-etching environments. This was attributed to the fact that scission of the C–S bond adjacent to the tertiary carbon atom produces a styryl radical with another styrene unit in the penultimate position, i.e. the radical should behave like a polystyrene radical with little or no tendency to depropagate.

Although the plasma and ion-milling resistance were excellent, the sensitivity was a factor of 10 less than that of the poly(olefin sulphone)s which may be attributed to the protective effect of the aromatic rings. This result is in accord with the lower $G(S)$ of poly(styrene sulphone) compared to that for the poly(olefin sulphone)s. The spun films also exhibited a strong tendency to crack during development when the film thickness was greater than 0·5 μm. Similar results were obtained for polymers containing higher ratios of styrene to SO$_2$, although the sensitivity decreased still further for ratios of styrene to SO$_2$ above 3:1.[56]

8.   X-RAY RESISTS

Poly(olefin sulphone)s also function as positive x-ray resists[88] as might be expected since the basic radiation chemistry is the same for both x-rays and electrons. Taylor[89] has reported a sensitivity of 94 mJ cm^{-2} for PBS exposed to Pd$_{L\alpha}$ x-rays ($\lambda = 4·37$ Å). Polysulphones are effective x-ray resists because of the presence of highly absorbing sulphur atoms (the mass absorption coefficient (μ_x) of sulphur at 4·37 Å is 1697 cm^2 g^{-1} compared to 100 cm^2 g^{-1} for carbon) which results in enhanced absorption relative to resists that contain mainly carbon atoms in the skeletal structure, e.g. poly(methyl methacrylate). However, sensitivity is still a factor of 10 too low for economic device fabrication and x-ray resist research has concentrated on the inherently more sensitive negative resists.[89] X-ray lithography is a technology which is still in its infancy and development of a highly sensitive positive resist remains a challenge to the chemist.

9. PHOTORESISTS

Poly(olefin sulphone)s do not absorb ultraviolet light in the region of the spectrum utilised by conventional photolithographic equipment, i.e. over the range 250–400 nm. Further, it is not possible to efficiently sensitise photodegradation of these materials, and consequently they have no utility as conventional near- or deep-u.v. photoresists. Although there has been no systematic study of the absorption spectra of aliphatic polysulphones, it is known that the sulphone group itself absorbs strongly below 200 nm, i.e. in the far-u.v. region. Since the photon energy at this wavelength is greater than the bond energy, it appeared likely that exposure of these materials to far-u.v. light would result in scission of the main chain.

Appelbaum and coworkers[90] obtained a sensitivity of $5 \, mJ \, cm^{-2}$ for PBS exposed at 185 nm using a pulsed xenon arc lamp. The advent of high-powered lasers covering this region of the spectrum has resulted in a resurgence of interest in far-u.v. lithography, particularly as some interesting ablative effects have been observed[91,92] at the high dose rates attainable with a laser source. The poly(olefin sulphone)s may yet prove useful in this application.

Bowden and Chandross[93] suggested that copolymers of arylated olefins, such as styrene, with SO_2 might undergo chain scission by absorption of u.v. light by the aromatic ring provided that there was efficient energy transfer from the absorbing chromophore to the C–S bond. They showed that the solubility of films of the 2:1 copolymer of styrene and SO_2 increased upon irradiation with a high-pressure mercury arc source, indicating that chain scission was the principal mode of degradation. The sensitivity ($18.4 \, J \, cm^{-2}$) was rather low because the absorbance of the benzene ring in styrene is weak ($\varepsilon \sim 200$ at 265 nm) and lies in a region where there is relatively little output from the high-pressure mercury arc lamp, but the use of high-powered lasers would overcome these disadvantages.

Bowden and Chandross attempted to increase the photosensitivity of this type of polymer by replacing some or all of the styrene units with larger aromatic hydrocarbons such as vinylbiphenyl, vinylnaphthalene or acenaphthylene, all of which have greater u.v. absorbance at somewhat longer wavelength. The most promising polymer from this class of compounds was a terpolymer based on styrene, acenaphthylene and SO_2, for which the sensitivity was about $500 \, mJ \, cm^{-2}$ for u.v. light in the wavelength region, 200–400 nm.

The photochemical reaction appeared to proceed most readily from the excited singlet state as triplet sensitisation, e.g. by benzophenone, was extremely inefficient. This may reflect the energy required to break the C–S bond, which is comparable to the triplet energy of the naphthylene group. Carbon atoms 1 and 9 of the fused-ring, acenaphthyl group are attached to the two carbon atoms in the main chain giving rise to a five-membered, strained ring and result in enhanced sensitivity relative to the related vinylnaphthalene polymer. These terpolymers formed brittle films similar to poly(styrene sulphone) which tended to crack during development. It was not possible to extend this synthetic approach to longer wavelengths because of the non-polymerisability of higher conjugated aryl systems.

An alternative approach towards development of a positive photoresist was reported by Himics and Ross.[94] The material was a copolymer of 5-hexene-2-one and SO_2 for which they reported a sensitivity of $520 \, mJ \, cm^{-2}$ for films sensitised with 30 % benzophenone and exposed to the full output of a 200-W mercury or mercury-xenon lamp. The degradation reaction was complicated by such factors as availability of oxygen, presence of sensitiser and precise wavelength range, suggesting several degradation pathways may be operable. The precise role of the pendant carbonyl was also not apparent.

10. CONCLUSIONS

It is clear that poly(olefin sulphone)s constitute an important class of resists applicable to all areas of lithography. These materials were first studied because of scientific interest in fundamental aspects of radiation degradation. Subsequent developments in the field of microelectronics were able to benefit substantially from such studies demonstrating the important synergism between fundamental and applied research. The lithographic versatility of the poly(olefin sulphone)s extends from conventional photolithography to the more esoteric forms of lithography either in use or under development. They have played a crucial role in the development of electron-beam lithography from a laboratory curiosity to a practical tool for manufacturing photolithographic masks. It is highly unlikely that the last word has been written on this interesting class of polymers.

REFERENCES

1. IVIN, K. J. and ROSE, J. B., *Adv. Macromol. Chem.*, **1** (1968), 335.
2. TOKURA, N., In: *Encyclopedia of Polymer Science and Technology*, Vol. 9, ed. H. Mark, New York, Interscience, 1968, p. 460.
3. FETTES, E. M. and DAVIS, F. O., In: *High Polymers*, Vol. 13, eds R. A. V. Raff and K. Doak, New York, Interscience, 1962, p. 225.
4. COLOMBO, P., FONTANA, J. and STEINBERG, M., *J. Polym. Sci., Part A-1*, **6** (1968), 3201.
5. BROWN, J. R. and O'DONNELL, J. H., *J. Polym. Sci., Part A-1*, **10** (1972), 1997.
6. NOSHAY, A. and PRICE, C. C., *J. Polym. Sci.*, **54** (1961), 533.
7. IVIN, K. J., LILLIE, E. D. and PETERSEN, I. H., *Makromol. Chem.*, **168** (1973), 217.
8. CORNO, C. and ROGGERO, A., *Europ. Polym. J.*, **12** (1976), 159.
9. FAWCETT, A. H., HEATLEY, F., IVIN, K. J., STEWART, C. D. and WATT, P., *Macromolecules*, **10** (1977), 765.
10. IVIN, K. J. and WALKER, N. A., *J. Polym. Sci., Part B*, **9** (1971), 901.
11. RYDEN, L. L. and MARVEL, C. S., *J. Am. Chem. Soc.*, **58** (1936), 2047.
12. OVENALL, D. W., SUDOL, R. S. and CABAT, G. A., *J. Polym. Sci., Polym. Chem. Ed.*, **11** (1973), 233.
13. CAIS, R. E. and O'DONNELL, J. H., *Macromolecules*, **9** (1976), 279.
14. CAIS, R. E., O'DONNELL, J. H. and BOVEY, F. A., *Macromolecules*, **10** (1977), 254.
15. CAIS, R. E. and STUK, G. J., *Macromolecules*, **13** (1980), 415.
16. CAIS, R. E. and STUK, G. J., *Polymer*, **19** (1978), 179.
17. HAZELL, J. E. and IVIN, K. J., *Trans. Faraday Soc.*, **58** (1962), 176.
18. HAZELL, J. E. and IVIN, K. J., *Trans Faraday Soc.*, **58** (1962), 342.
19. HAZELL, J. E. and IVIN, K. J., *Trans Faraday Soc.*, **61** (1965), 2330.
20. GIPSTEIN, E., MOREAU, W., CHIU, G. and NEED, O. U., *J. Appl. Polym. Sci.*, **21** (1977), 677.
21. STILLE, J. K. and THOMSON, D. W., *J. Polym. Sci.*, **62** (1962), S 118.
22. STILLE, J. K. and EMPEN, J. A., In: *The Chemistry of Sulfides*, ed. A. V. Tobolsky, New York, Interscience, 1968, p. 125.
23. FOLDI, V. S. and SWEENY, W., *Makromol. Chem.*, **72** (1964), 208.
24. MARVEL, C. S. and ALDRICH, P. H., *J. Am. Chem. Soc.*, **75** (1953), 1997.
25. WELLISCH, E., GIPSTEIN, E. and SWEETING, O. J., *J. Appl. Polym. Sci.*, **8** (1964), 1623.
26. FOSTER, F. C., *J. Am. Chem. Soc.*, **74** (1952), 2299.
27. BROWN, J. R. and O'DONNELL, J. H., *J. Appl. Polym. Sci.*, **23** (1979), 2763.
28. DAINTON, F. S. and IVIN, K. J., *Quart. Rev.*, **12** (1958), 61.
29. BUSFIELD, W. K., In: *Aspects of Degradation and Stabilization of Polymers*, ed. H. H. G. Jellinek, Amsterdam, Elsevier, 1978.
30. DAINTON, F. S. and IVIN, K. J., *Proc. Roy. Soc. (London)*, **A212** (1952), 217.
31. AYSCOUGH, P. B., IVIN, K. J. and O'DONNELL, J. H., *Trans. Faraday Soc.*, **61** (1965), 1601.
32. COOK, R. E., IVIN, K. J. and O'DONNELL, J. H., *Trans. Faraday Soc.*, **61** (1965), 1887.

33. BRADY, B. H. G. and O'DONNELL, J. H., *Trans. Faraday Soc.*, **64** (1968), 23.
34. BRADY, B. H. G. and O'DONNELL, J. H., *Trans. Faraday Soc.*, **64** (1968), 29.
35. SNOW, R. D. and FREY, F. E., *J. Am. Chem. Soc.*, **65** (1943), 2417.
36. DAINTON, F. S. and IVIN, K. J., *Nature (London)*, **162** (1948), 705.
37. COOK, R. E., DAINTON, F. S. and IVIN, K. J., *J. Polym. Sci.*, **26** (1957), 351.
38. COOK, R. E., DAINTON, F. S. and IVIN, K. J., *J. Polym. Sci.*, **29** (1958), 549.
39. BRAY, B. G., *Diss. Abst.*, **19** (1958), 494.
40. EATON, E. C. and IVIN, K. J., *Polymer*, **6** (1965), 339.
41. BROWN, J. R. and O'DONNELL, J. H., *Macromolecules*, **3** (1970), 265.
42. BROWN, J. R. and O'DONNELL, J. H., *Macromolecules*, **5** (1972), 109.
43. O'DONNELL, J. H., SMITH, C. A. and WINSOR, D. J., *J. Polym. Sci., Polym. Phys. Ed.*, **16** (1978), 1515.
44. O'DONNELL, J. H., RAHMAN, N. P., SMITH, C. A. and WINSOR, D. J., *Macromolecules*, **12** (1979), 113.
45. NICHOL, J. M., O'DONNELL, J. H., RAHMAN, N. P. and WINSOR, D. J., *J. Polym. Sci., Polym. Chem. Ed.*, **15** (1977), 2919.
46. JAMES, P. M., OUANO, A. C., GIPSTEIN, E. and GREGGES, A. R., *J. Appl. Polym. Sci.*, **16** (1972), 2425.
47. GRAY, D. N., *Polym. Eng. Sci.*, **17** (1977), 719.
48. CRAWFORD, J. E. and GRAY, D. N., *J. Appl. Polym. Sci.*, **15** (1971), 1881.
49. O'DONNELL, J. H., PhD Thesis, University of Leeds, 1962.
50. BOWDEN, M. J. and THOMPSON, L. F., *J. Appl. Polym. Sci.*, **17** (1973), 3211.
51. BOWDEN, M. J., *J. Polym. Sci., Polym. Chem. Ed.*, **12** (1974), 499.
52. BOWMER, T. N. and O'DONNELL, J. H., *J. Macromol. Sci., Chem.*, **A-17** (1982), 243.
53. BOWMER, T. N., O'DONNELL, J. H. and WELLS, P. R., *Polym. Bull.*, **2** (1980), 103.
54. BOWMER, T. N., O'DONNELL, J. H. and WELLS, P. R., *Makromol. Chem., Rapid Commun.*, **1** (1980), 1.
55. BOWMER, T. N., O'DONNELL, J. H., *J. Polym. Sci., Polym. Chem. Ed.*, **19** (1981), 45.
56. STILLWAGON, L. E., DOERRIES, E. M., THOMPSON, L. F. and BOWDEN, M. J., *Coatings and Plastics Preprints*, **37**(2) (1977), 38.
57. CAIS, R. E. and O'DONNELL, J. H., *Makromol. Chem.*, **176** (1975), 3517.
58. DOUBE, C. P., Honours Thesis, University of Queensland, 1970.
59. CAIS, R. E. and O'DONNELL, J. H., *J. Polym. Sci., Polym. Lett. Ed.*, **14** (1976), 263.
60. JOHNSON, R. N., *Encyclopedia of Polymer Sci.*, **11** (1969), 447.
61. HILL, E. H. and CALDWELL, J. R., *J. Polym. Sci., A*, **2** (1969), 1251.
62. BOWDEN, M. J., THOMPSON, L. F., ROBINSON, W. and BIOLSI, M., *Macromolecules*, **15** (1982), 1417.
63. BOWMER, T. N. and O'DONNELL, J. H., *Polymer*, **22** (1981), 71.
64. THOMPSON, L. F., WILLSON, C. G. and BOWDEN, M. J. (Eds), *Introduction of Microlithography*, ACS Symposium Series No. 219, Washington DC, American Chemical Society, 1983.
65. BOWDEN, M. J. and THOMPSON, L. F., *J. Electrochem. Soc.*, **120** (1973), 1722.
66. BOWDEN, M. J. and THOMPSON, L. F., *J. Appl. Polym. Sci., Polym. Symp.*, **23** (1974), 99.

67. BOWDEN, M. J., THOMPSON, L. F. and BALLANTYNE, J. P., *J. Vac. Sci. Technol.*, **12** (1975), 1294.
68. BOWDEN, M. J., *J. Polym. Sci., Polym. Symp.*, **49** (1975), 221.
69. KU, H. Y. and SCALA, L. C., *J. Electrochem. Soc.*, **116** (1969), 980.
70. INOKUTI, M., *J. Chem. Phys.*, **38** (1963), 1174.
71. CHU, W. H., GIPSTEIN, E. and OUANO, A. C., *J. App. Polym. Sci.*, **21** (1977), 1045.
72. GIPSTEIN, E. and HEWETT, W. A., US Patent 3 398 350, 1975.
73. HIMICS, R. J., DESAI, N., KAPLAN, M. and POLINIAK, E. S., *ACS Div. Org. Coatings and Plast. Chem. Preprints*, **35**(2) (1975), 266.
74. HIMICS, R. J., KAPLAN, M., DESAI, N. and POLINIAK, E. S., *ACS Div. Org. Coatings and Plast. Chem. Preprints*, **35**(2) (1975), 273.
75. POLINIAK, E. S., SCHEIBLE, H. G. and HIMICS, R. J., US Patent 3 935 331, 1976.
76. BOWDEN, M. J. and THOMPSON, L. F., *J. Electrochem. Soc.*, **121** (1974), 1620.
77. BOWDEN, M. J. and THOMPSON, L. F., *Proc. 6th. Intl. Conf. on Electron and Ion Beam Sci. and Technol.*, ed. R. Bakish, Princeton, New Jersey, Electrochem. Soc., 1974, p. 81.
78. BOWDEN, M. J. and THOMPSON, L. F., *Polymer Eng. and Sci.*, **17** (1977), 269.
79. BOWDEN, M. J. and THOMPSON, L. F., *Polymer Eng. and Sci.*, **14** (1974), 525.
80. BOWMER, T. N. and BOWDEN, M. J., *Org. Coatings and App. Polym. Sci. Proc.*, **48** (1983), 161.
81. BOWDEN, M. J., PEASE, R. F. W., YAU, L. D., FRACKOVIAK, J., THOMPSON, L. F., SKINNER, J. G. and BALLANTYNE, J. P., In: *Microcircuit Engineering*, eds H. Ahmed and W. C. Nixon, Cambridge, Cambridge University Press, 1980, p. 239.
82. HIMICS, R. J., DESAI, N. V. and POLINIAK, E. S., US Patent 4 045 318, 1977.
83. YAMAZAKI, T., WATAKABE, Y., SUZUKI, Y. and NAKATA, H., *J. Electrochem. Soc.*, **127** (1980), 1859.
84. BOWDEN, M. J., THOMPSON, L. F., FAHRENHOLTZ, S. R. and DOERRIES, E. M., *J. Electrochem. Soc.*, **128** (1981), 1304.
85. BOWDEN, M. J. and THOMPSON, L. F., US Patent 4 289 845, 1981.
86. SHIRAISHI, H., ISOBE, A., MURAI, F. and NONOGAKI, S., *Org. Coatings and Appl. Polym. Sci. Proc.*, **48** (1983), 178.
87. CHENG, Y. Y., GRANT, B. D., PEDERSON, L. A. and WILLSON, C. G., US Patent 4 398 001, 1983.
88. THOMPSON, L. F., FEIT, E. D., BOWDEN, M. J., LENZO, P. V. and SPENCER, E. G., *J. Electrochem. Soc.*, **121** (1974), 1500.
89. TAYLOR, G. N., *Solid State Technol.*, **23** (1980), 73.
90. APPELBAUM, J., BOWDEN, M. J., CHANDROSS, E. A., FELDMAN, M. and WHITE, D. L., *Proc. Kodak Microelectronic Seminar—Interface 1975*, 19–21 October 1975, p. 40.
91. SRINIVASAN, R. and MAYNE-BANTON, V., *Appl. Physics Lett.*, **41** (1982), 576.
92. GEIS, M. W., RANDALL, J. N., DEUTSCH, T. F., EFREMOW, N. N., DONNELLY, J. P. and WOODHOUSE, D., *J. Vac. Sci. Technol. B*, **1** (1983), 1178.
93. BOWDEN, M. J. and CHANDROSS, E. A., *J. Electrochem. Soc.*, **122** (1975), 1370.
94. HIMICS, R. J. and ROSS, D. L., In: *Proc. Reg. Tech. Conf. Photopolymers: Principles, Processes and Materials*, 13–15 October 1975, Ellenville, New York, Mid-Hudson Sect., SPE, 1976, p. 26.

Chapter 3

POLYMER STABILITY IN AGGRESSIVE MEDIA

G. E. ZAIKOV

*Institute of Chemical Physics, Academy of Sciences of the USSR,
Moscow, USSR*

SUMMARY

This chapter deals with the kinetic laws and the mechanism of hydrolytic degradation of polymers, particularly the heterochain polymers or those containing heteroatoms in the side groups. The following problems are discussed: catalysis in aggressive liquid media, such as solutions of acids, bases and salts; the reactivity and mechanisms of transformation of chemically unstable bonds in various polymers; specific features of the chemical degradation of polymers compared with that of low-molecular-compounds; diffusion of aggressive media in polymers; prediction of polymer stability in aggressive media; the influence of the medium on the mechanical properties of polymers. Examples are given of the use of hydrolytic degradation as a method of modification of polymeric products.

1. INTRODUCTION

The chemical, including the hydrolytic, degradation of polymers that takes place when polymers come into contact with aggressive media, is a complex physical and chemical process involving the diffusion of the aggressive medium in the polymer and the subsequent reactions of chemically unstable bonds.

To establish quantitative kinetic laws and mechanisms of polymer degradation in aggressive media the following must be elucidated:[1-4]

1.   The mechanism of transformation of the chemically unstable bonds of various polymers in the stated media.
2.   Types of macromolecular decomposition and the specific features of the polymeric state compared with the low-molecular state.
3.   The transfer laws for the aggressive media in polymeric matrices.

Knowledge of the mechanisms of degradation of polymers in aggressive media allows one to predict the end-use properties of polymeric products in contact with these media, as well as to find ways of increasing the chemical stability of the polymers.[1-4] Aggressive media are substances which, under specific conditions, cause a change in the chemical structure of the polymer and, as a result, a change in the properties of the polymeric material.

This chapter will deal with the hydrolytic stability of polymers in aggressive media, acids and solutions of bases and salts. Oxidative degradation processes will not be considered.

Practically all polymers are unstable towards substances that possess oxidative properties. However, it is for the most part polymers with heteroatoms in the main or side chain that are unstable towards acids and alkaline and salt solutions. Carbon chain polymers with no double bonds in their backbones and no heteroatoms in their side chains are theoretically stable.

2.   REACTIVITY OF CHEMICALLY UNSTABLE BONDS AND THE MECHANISM OF THEIR TRANSFORMATIONS

In investigating polymer degradation in aggressive media it is very important to know the transformation mechanism of the chemically unstable bonds and the values of the parameters of the kinetic equations. In this section attention is drawn to the transformations in aggressive media of model compounds with various unstable bonds.

2.1. Compounds with Amido and Imido Bonds

2.1.1. Acid-catalysed Decomposition of Amides
For most amides it is amido-bond decomposition that is observed in acid hydrolysis or solvolysis, though some amides can decompose at the alkyl–nitrogen bond.[5]

Protonation of amides. The electronic structure of the amido-bond predetermines two possible ways in which protonation may occur, namely at the carbonyl oxygen and the amido nitrogen:[6-9]

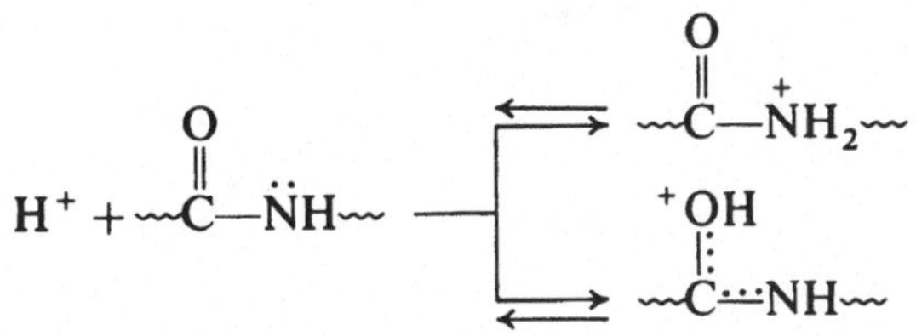

Molecularity of the rate determining step. The rate determining step of hydrolysis is bimolecular. Within the framework of the bimolecular mechanism the possible pathways of hydrolysis can be represented by the following scheme:[10-12]

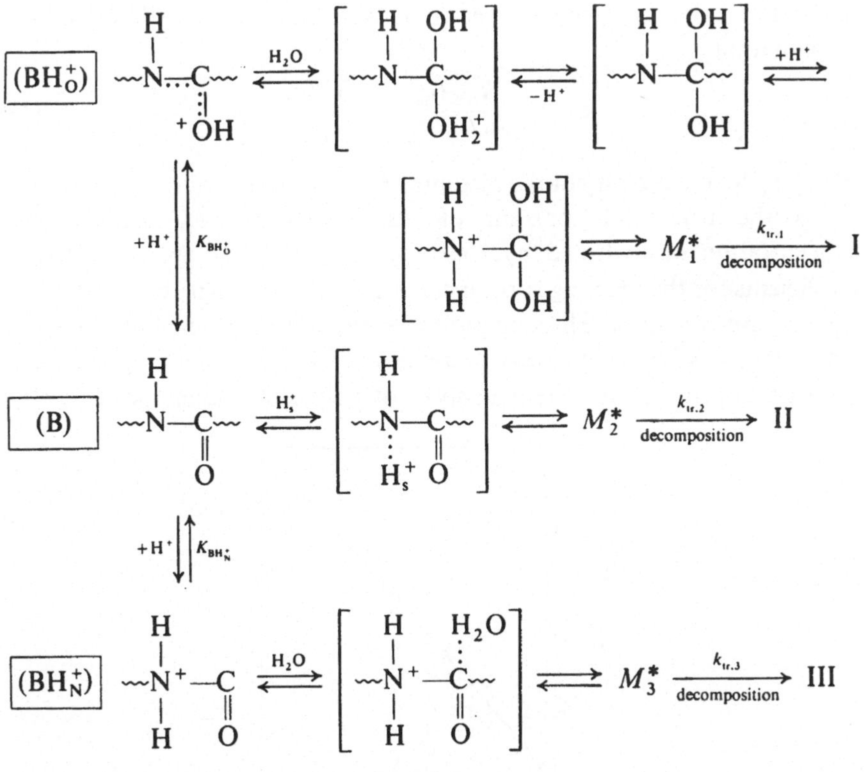

Strictly speaking, hydrolysis can follow any of the three pathways.[1-3,10-12] However, the decomposition of the majority of amides in acid media occurs by mechanism A-2 (acid catalysed, bimolecular)

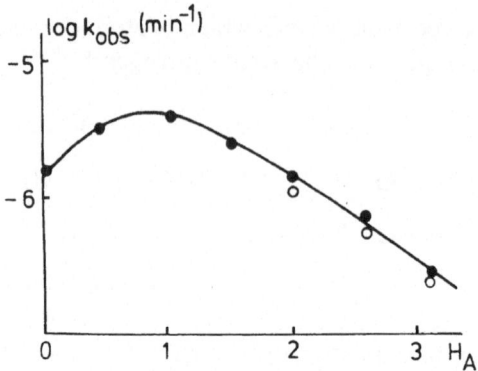

FIG. 1. Dependence of $\log k_{obs}$ (25 °C) on the acidity of aqueous solutions of sulphuric acid for the hydrolysis of polycaproamide cyclic trimer (●) and polycaproamide with $\bar{M}_v = 20,000$ (○).[10]

(pathway II) and the effective kinetic constant is satisfactorily described by the equation

$$k_{obs} = \frac{k_{tr} \cdot H_s^+}{1 + h_A/K_{BH_O^+}} \cdot \frac{f_B \cdot f_{H_s^+}}{f^*} \tag{1}$$

where h_A denotes acidity of the medium, $K_{BH_O^+}$ is the equilibrium constant, k_{tr} is the true rate constant of decomposition, H_s^+ denotes the concentration of hydrated hydrogen ions, f_B and $f_{H_s^+}$ are the activity coefficients for the reacting substances and f^* is the activity coefficient for the activated complex. This equation describes the experimental results of acid hydrolysis for the majority of amides.[1-3] For example, Fig. 1 shows the variation in k_{obs} for the hydrolysis of polycaproamide and its cyclic

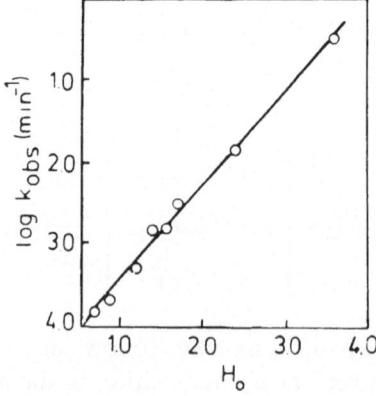

FIG. 2. Dependence of $\log k_{obs}$ (25 °C) on H_0 for the hydrolysis of dinitro-2-pyridylalanylglycyl in aqueous solutions of sulphuric acid.[11]

trimer versus the acidity function H_A ($= -\log k_a$) in aqueous solutions of sulphuric acid.[10] It is of practical importance that, for compounds with an amido bond, k_{obs} passes through a maximum with respect to the acid concentration, and the hydrolytic stability of these compounds increases in concentrated solutions of acids. This is associated with the competition between protonation at carbonyl oxygen and amido nitrogen. Protonation at nitrogen prevails in less concentrated acids and results in reaction, while protonation at oxygen prevails in more concentrated acids, this form being non-reactive. For a small number of amides (e.g. for amides in concentrated $H_2SO_4 > 70\%$) the decomposition in acids proceeds by a monomolecular mechanism:[11]

and the variation in k_{obs} is described by the equation (Fig. 2)

$$k_{obs} = \frac{k_{tr}}{K_{BH_N^+}} \cdot h_0 \qquad (2)$$

2.1.2. Base-catalysed Decomposition of Amides

As in acid decomposition, the majority of amides decompose at the amido bond in basic media,[1-3] although there is some evidence for amide decomposition at the alkyl–nitrogen bond.[13]

Ionisation of amides. In basic media amides can become ionised by attaching the hydroxyl ion to the carbonyl group or by detaching a hydrogen atom from the amido nitrogen. The latter pathway is ruled out for the tertiary amides:[14-16]

Molecularity of the rate determining step. The scheme for the decomposition of amides can be represented as follows:[1]

Pathways I and II are the most probable.[1,2] For the general case the equation for the kinetic constant has been obtained in following form

$$k_{obs} = \frac{k_{trBOH^-} \cdot K_{BOO^=} \cdot \dfrac{a_{H_2O}^2 f_{BOH^-}}{b_0} \dfrac{f_{BOO^=}}{f_1^*} + k_{trBOO^=} \cdot a_{H_2O} \dfrac{f_{BOO^=}}{f_2^*}}{1 + K_{BOO^=} \cdot \dfrac{a_{H_2O}}{b_0} + K_{BOH^-} \cdot K_{BOO^=} \cdot \dfrac{a_{H_2O}}{b_0}} \tag{3}$$

where K_{BOH^-} and $K_{BOO^=}$ are the equilibrium constants for the formation of BOH^- and $BOO^=$, a_{H_2O} denotes the activity of the water, b_0 is the basicity of the medium, and f denotes the respective activity coefficients of the various particles. The precise mechanism of the participation of water in the rate determining step is unknown for pathways I and II. Two possibilities are suggested, namely that the rate determining step is either the interaction between the ionised form of the reagent and the water molecule or it is the decomposition or ionisation of the complexes formed by the ionised forms and water.

TABLE 1

KINETIC PARAMETERS FOR N-METHYLACETAMIDE HYDROLYSIS IN AQUEOUS SOLUTIONS
OF POTASSIUM HYDROXIDE

KOH (mass %)	$-\log k_{obs}$ $(min^{-1}, 25°C)$	$-b_0$	$-\log a_{H_2O}$	$-\log \dfrac{k_{obs}}{b_0 a_{H_2O}}$
2·00	3·80	−0·56	0·00	3·24
4·61	3·46	−0·20	0·01	3·25
8·99	3·11	0·17	0·02	3·26
13·89	2·82	0·47	0·04	3·25
18·47	2·60	0·72	0·07	3·25

Particular cases of eqn (3) will be considered.

Case 1. The concentration of B is much higher than that of BOH$^-$ and BOH$^=$. Hence:[15]

$$k_{obs} = (k_{trBOH^-}/K_{BOH^-})b_0 a_{H_2O} \frac{f_{BOH^-}}{f_1^*} \qquad (4)$$

This condition is realised for N-methylacetamide (Table 1) in non-concentrated alkaline solutions (Fig. 3).

Case 2. The concentrations of B and BOH$^-$ are similar, while the concentration of BOO$^=$ is small. Then:

$$k_{obs} = \frac{k_{trBOH^-} a_{H_2O}}{1 + K_{BOH^-}/b_0} \cdot \frac{f_{BOH^-}}{f_1^*} \qquad (5)$$

For N-methylacetamide[15] this corresponds to alkaline concentrations in the range 18–44%.

Case 3. The concentration of BOO$^=$ is much higher than those of BOH$^-$ and B. As a result:

$$k_{obs} = k_{trBOO^=} a_{H_2O} \frac{f_{BOO^=}}{f_2^*} \qquad (6)$$

A similar equation is derived if the concentration of BOH$^-$ is much higher than that of BOH$^=$ and B. A graphic solution of eqns (4) and (6) for N-methylacetamide hydrolysis is given in Fig. 3.

Thus, the decomposition process of most amides proceeds via mechanism B-2 and is described by eqn (3).

2.1.3. Peculiarities of Imido-bond Decomposition

For the cleavage of a polymeric chain of polyamide, the cleavage of two bonds, i.e. imido and o-carboxy amido, is required. N-phenylphthalimide (I) and N-phenylphthalamino acid (II) are simple model compounds containing such bonds

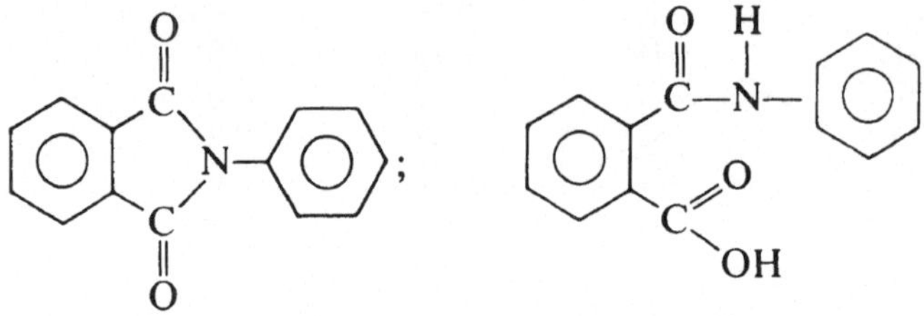

Acid-catalysed decomposition. At present the following scheme for o-carboxyamide decomposition is accepted:[16,17]

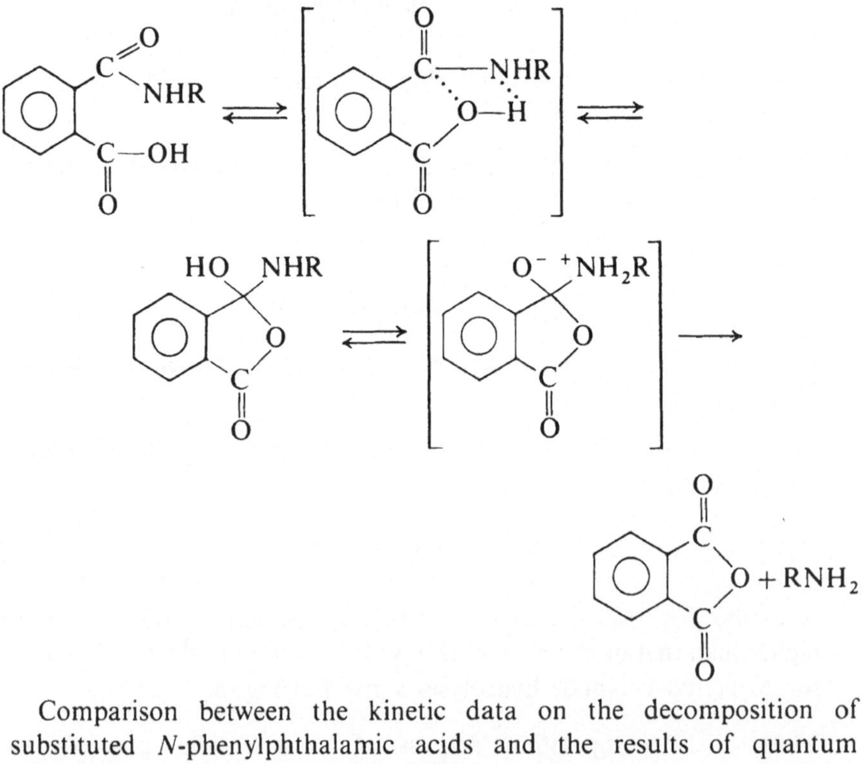

Comparison between the kinetic data on the decomposition of substituted N-phenylphthalamic acids and the results of quantum mechanical calculations[17] shows that o-carboxyamide decomposition

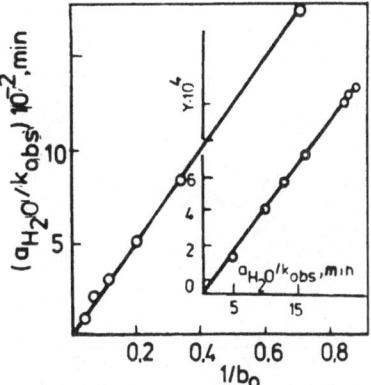

FIG. 3.   Graphic determination of $k_{tr(BOH^-)}$, $k_{tr(BOH^=)}$, K_{BOH^-} and $K_{BOO^=}$ from eqns (4) and (6) for the hydrolysis of N-methylacetamide.[15]

$$y = - K_{BOH^-} \cdot (a_{H_2O}/b_0)((-1/b_0) + (k_{tr(BOH^-)}/K_{BOH^-})(a_{H_2O}/k_{obs})) + (a_{H_2O}/b_0)$$

passes through an equilibrium transition of isomer **A** (an energetically stable form) to isomer **B** (a reactive form in the decomposition reaction).

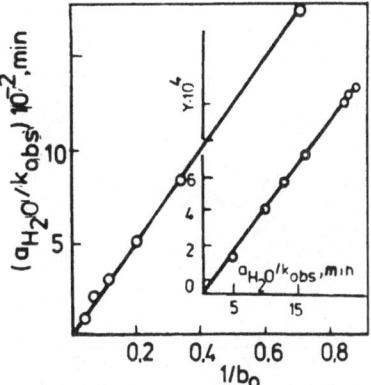

The acid-catalysed decomposition proceeds along two pathways:[1-3,17] in the range of moderately concentrated acid solutions ($H_A > 1$) an intramolecular acid catalysis takes place (Fig. 4) (the non-ionised and non-protonated form is reactive), while in the range of concentrated solutions ($H_A < 1$) the acid catalysis is effected by external solvated protons. The variation in k_{obs} is described by eqn (7):

$$k_{obs} = \frac{k_{tr} H_s^+ + k_{tr}}{1 + h_A/K_{BH_O^+} + K_{dis}/h_A} \tag{7}$$

where k_{tr} is the true constant of the intramolecular decomposition rate; K_{dis} is the decomposition constant of the carboxyl group.

The acid-catalysed decomposition of N-phenylphthalimide (similar to

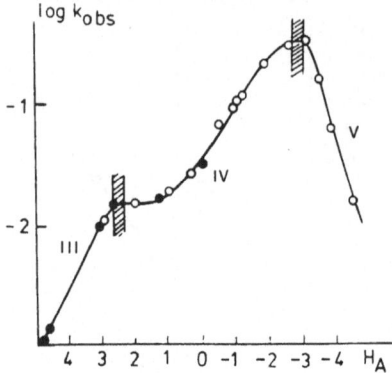

FIG. 4. Dependence of $\log k_{obs}$ (25 °C) on the acidity function of aqueous solutions of sulphuric acid for the degradation reaction of N-phenylphthalamino acid: (●) according to data from ref. 17; (○) according to data from ref. 1.

This shows the boundaries of the regions where forms **III**, **IV** and **V** of the hydrolysed substances predominate.

many other imides) proceeds by mechanism A-2. The variation in k_{obs} is described by eqn (8):[1-3]

$$k_{obs} = k_{tr} H_s^+ \text{ const} \tag{8}$$

Base-catalysed decomposition. Decomposition of N-phenylphthalamino acid in alkali proceeds by mechanism B-2 (base catalysed, bimolecular) and is described by eqn (3). The variation in k_{obs} for N-phenylphthalamide decomposition under these conditions is described[18] by eqn (9):

$$k_{obs} = k_0 + \frac{k_{tr}}{k_{BOH^-}} b_0 a_{H_2O} \tag{9}$$

where k_0 is the rate constant for a non-catalytic process.

Comparison between the reactivities of N-phenylphthalamino acid and N-phenylphthalamide in acid and basic media is presented in Fig. 5. It is seen that basic media offer conditions favourable for the conversion of imide to its intermediate product, the amido acid; while different reactivities of imido and non-cyclised o-carboxyamido bonds in acid media

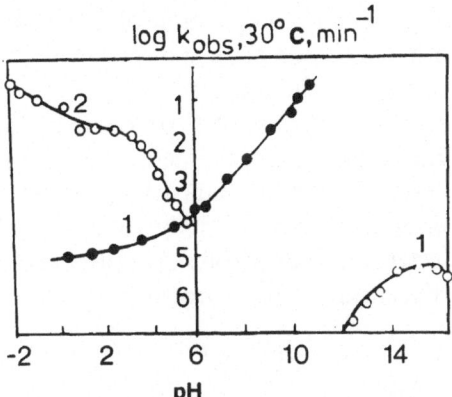

FIG. 5.   Comparison between the reactivities of o-carboxy-amido (1) and -imido (2) bonds for various values of pH.[17,18]

have made it possible to develop an absolute kinetic method for determining the degree of cyclisation of soluble imides.[18]

Papers[1-3,19-22] have presented the corresponding kinetic equations and considered the mechanisms of hydrolytic decomposition for other nitrogen-containing model compounds, including benzanilides, benzoxazoles, benzimidazoles, etc.

2.2. Compounds with Ester Bonds
Similar to amides, compounds with an ester bond decompose in both acidic and basic media.

2.2.1. Acid-catalysed Decomposition of Esters
It has been suggested,[23-25] that the decomposition of esters proceeds by three mechanisms:

$$B \ldots H_2SO_4 \quad M_1^* \xrightarrow{k_{tr.1}} \text{I}$$
$$BH^+ \underset{}{\overset{+H_2O}{\rightleftarrows}} M_2^* \xrightarrow{k_{tr.2}} \text{II}$$
$$B \ldots H^+(OH_2)_n \xrightarrow{K_n} M_3^* \xrightarrow{k_{tr.3}} \text{III}$$

Since the structure of the protonated form of the ester is unknown, it has been denoted by BH^+ in the scheme. The protonated form of an ester can form non-reactive particles with bases, acid anions and H_2O molecules present in the solution. The decomposition of BH^+ (mechanism A_{Ac}-1) and that of BH^+ with one or two water molecules (mechanism B_{Ac}-2) are the rate determining steps of the process.

The acid hydrolysis of esters can proceed by several mechanisms,[11,23-26] described by complex kinetic equations. However, the decomposition of ester bonds in industrial polyesters proceeds via mechanism A_{Ac}-2 in dilute and moderately concentrated acid solutions. For the majority of esters k_{obs} (of hydrolysis) in these solutions increases in proportion to the concentration of hydrated protons.

$$k_{obs} = \frac{k_{tr,H_s^+}H_s^+ + k_{tr,H_s^+Y} H_s^+Y}{1 + h_0/K_{BH^+} + a_{H_s^+}/K_h} \tag{10}$$

Equation (10) provides a good explanation of the effect of neutral salts and other particles Y (non-dissociated acids, etc.) on the rate of the reaction. (K_h is the equilibrium constant in the formation of a complex with the hydrated proton BH_s^+.)

It is believed[11] that the form protonated at the alkyl oxygen is reactive, and that the reaction occurs by the following sequence:

$$\underset{R-O-C\sim}{\overset{O}{\parallel}} \underset{\xrightleftharpoons{H^+}}{} \underset{R-O-C\sim}{\overset{H^+\ O}{\underset{|}{}\ \parallel}} \underset{\xrightleftharpoons{M^+}}{} R^+ + \underset{HO-C\sim}{\overset{O}{\parallel}}$$

2.2.2. Base-catalysed Decomposition of Esters

Decomposition of esters in basic media occurs very rapidly; the kinetics of the reactions are therefore difficult to investigate. It has been established that hydrolysis proceeds[11] through the formation of a tetrahedral intermediate compound:

$$\underset{\sim C-O-R + B^-}{\overset{O}{\parallel}} \underset{\xrightleftharpoons[k_{-1}]{k_1}}{} \underset{\underset{B}{\overset{O^-}{\underset{|}{\sim C-O-R}}}}{} \xrightleftharpoons{+H_2O} M^* \xrightarrow[\text{decomposition}]{k_{tr}}$$

the presence of which has been identified.[1-3] In general, k_{obs} for ester hydrolysis in basic media is given by

$$k_{obs} = k_0 + \sum_i (k_B c_B)i \tag{11}$$

where k_0 and k_B are the effective rate constants for ester decomposition in water and bases respectively. The decomposition of esters in basic media proceeds by mechanisms B_{Ac}-2 and B_{A1}-2.[2]

2.3. Compounds with Acetal Bonds

The decomposition of compounds with acetal bonds is only catalysed by acids. Acetals decompose 7–8 orders faster than ethers.[27] The higher reactivity of acetals, compared to that of other oxygen-containing compounds, is associated with the induction effect of two neighbouring oxygen atoms:

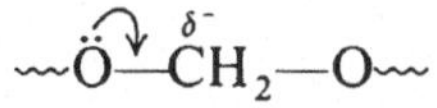

$$\sim\overset{..}{O}\overset{\delta^-}{-}\dot{C}H_2-O\sim$$

Acetal hydrolysis proceeds via mechanism A-1 and is described by eqn (12):

$$k_{obs} = \frac{k_{tr}}{K_{BH^+}} \cdot h_0 \cdot \frac{f_{BH^+}}{f^*} \qquad (12)$$

2.3.1. Decomposition of Glycosides

Glycosides are a variety of cyclic acetals:

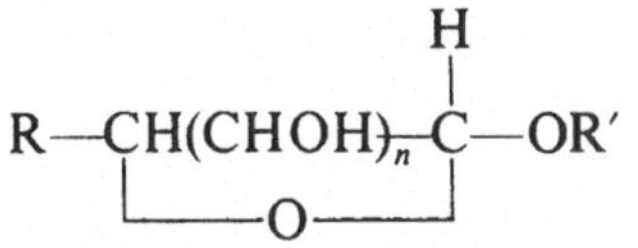

$$\text{R—CH(CHOH)}_n\text{—C—OR'}$$

Cyclic[28] and acyclic[29] mechanisms have been used to describe the hydrolysis of these compounds. The cyclic mechanism, accepted by most authors, includes the following steps:

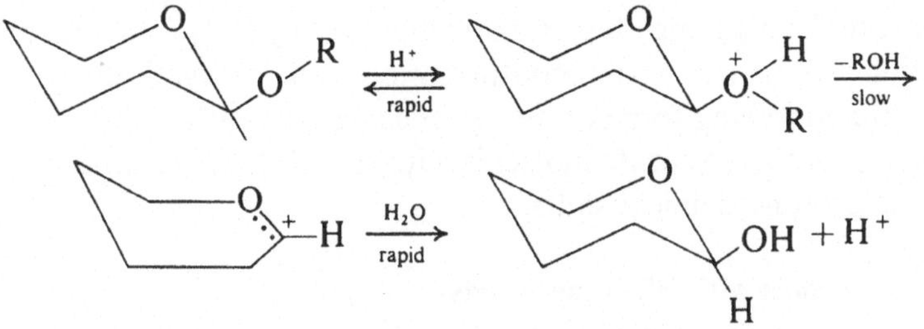

The process involves a fast equilibrium protonation of the glycoside oxygen and a slow heterolysis with the formation of a cyclic carbonium ion that exists in the semi-chair conformation.

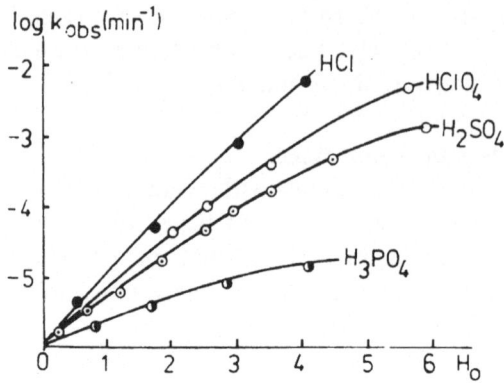

FIG. 6. Dependence of $\log k_{obs}$ (25 °C) on the acidity function H_o in aqueous solutions of hydrochloric (●), perchloric (○), sulphuric (◉), and phosphoric (◑) acids for the hydrolysis of cellobiose.[30]

Mechanism A-1 is commonly accepted for the hydrolysis reaction of glycosides[30] (Fig. 6). The slope of the curves in Fig. 6 is less than unity and varies with the acid concentration. The hydrolysis of pyranosides is essentially affected by the interaction of the glycoside molecule with the medium through the hydroxyl group at C2. Equation (13) satisfactorily describes the experimental facts of the acid-catalysed hydrolysis of cellobiose and other glycosides:

$$\log \frac{k_{obs}}{h_0} = \rho_{\Delta\delta} + \log \frac{k_{tr}}{K_{BH^+}} \tag{13}$$

where $\Delta\delta$ denotes the chemical shift of protons in p.m.r. spectra with a change of medium, and ρ is the proportionality factor.

The rate determining step in the cellobiose hydrolysis reaction is the transition from the chair conformation (an energetically favoured carbonium ion being formed), this step being especially influenced by the interaction of the glycoside molecule with the medium which is evaluated from the proton chemical shifts.

2.4. Compounds with Siloxane Bonds

At present there are several views about the mechanism of acid catalysed decomposition of organosiloxanes.[31−33] It is suggested that splitting proceeds by an S_N1 mechanism, i.e. a rapid and equilibrium protonation of oxygen takes place in the siloxane bond with the subsequent slow decomposition of the protonated form into silanol and the silyl cation, the latter reacting rapidly with a nucleophilic reagent, e.g. alcohol:

$$\equiv Si-O-Si\equiv \underset{}{\overset{H^+}{\rightleftharpoons}} \equiv Si-\overset{\overset{\displaystyle H}{|}}{\underset{+}{O}}-Si\equiv \qquad \text{(rapidly)}$$

$$\equiv Si-\overset{\overset{\displaystyle H}{|}}{\underset{+}{O}}-Si\equiv \rightleftharpoons \equiv Si-OH + \overset{+}{Si}\equiv \qquad \text{(slowly)}$$

$$\equiv \overset{+}{Si} + HOR \rightleftharpoons SiOR + H^+ \qquad \text{(rapidly)}$$

A bimolecular S_N2 mechanism has also been suggested[31] according to which the interaction of the nucleophilic reagent with the protonated form of organosiloxane is the limiting step.

$$\equiv Si-O-Si\equiv \underset{}{\overset{H^+}{\rightleftharpoons}} \equiv Si-\overset{\overset{\displaystyle H}{|}}{\underset{+}{O}}-Si\equiv \qquad \text{(rapidly)}$$

$$\equiv Si-\overset{\overset{\displaystyle H}{|}}{\underset{+}{O}}-Si\equiv + ROH \rightleftharpoons \equiv Si-OH + \equiv Si-\overset{\overset{\displaystyle H}{|}}{\underset{+}{O}}R \qquad \text{(slowly)}$$

$$\equiv Si-\overset{\overset{\displaystyle H}{|}}{\underset{+}{O}}-R \rightleftharpoons \equiv SiOR + H^+ \qquad \text{(rapidly)}$$

The kinetics of decomposition of octamethylcyclotetrasiloxane (D_4) has been subjected to a particularly detailed investigation.[32-33] The D_4 decomposition reaction was conducted under intensive stirring, the conditions ensuring independence of the reaction rate from the intensity of stirring. The decomposition constant k_{obs} was estimated from the equation:

$$-\ln\frac{m}{m_0} = k_{obs}\frac{S_0}{m_0}t \qquad (14)$$

where m_0 and m stand for the masses at times t_0 and t and S_0 denotes the total surface of D_4 drops.

Alcoholysis of cyclic and linear organosiloxanes follows an S_N2 mechanism,[32,33] the nature of the alcohol used having a predominating influence on the process.

Thus, analysis of the literature data makes it possible to reveal two

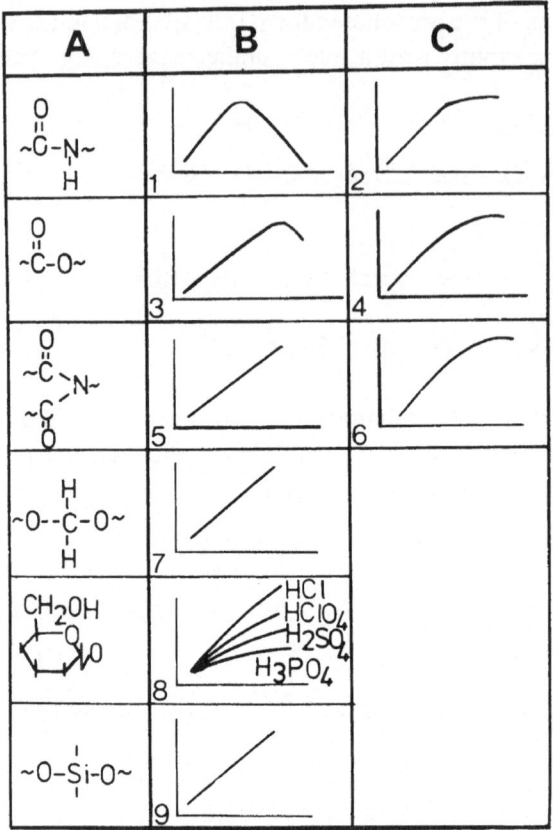

Fig. 7.   Decomposition of major chemical bonds. (A) Chemical bonds. (B) Acid catalysis and (C) base catalysis ($\log k_{obs}$ (ordinate) versus % acid/base (abscissa)). 1. $E \simeq 20\,\text{kcal mol}^{-1}$;   2. $E \simeq 16\,\text{kcal mol}^{-1}$;   3. $E \simeq 18\,\text{kcal mol}^{-1}$;   4. $E \simeq 14\,\text{kcal mol}^{-1}$; 5. $E \simeq 20\,\text{kcal mol}^{-1}$; 6. $E \simeq 16\,\text{kcal mol}^{-1}$; 7. $E \simeq 24\,\text{kcal mol}^{-1}$; 9. $E \simeq 12\,\text{kcal mol}^{-1}$.

regularities in the decomposition of compounds with chemically unstable bonds in aggressive media. If it is the non-ionised form of the compound that is reactive (decomposition of carbonyl-containing compounds by mechanism A-2), carbonyl-group protonation decreases the concentration of the reactive form, resulting in a considerable decrease in k_{obs} (increased chemical stability) in the media. If it is the ionised form that is reactive (decomposition of compounds by mechanisms A-1 and B-2), k_{obs} increases, as a rule, with the increase in concentration of the aggressive medium, with the exception of a number of reactions proceeding by mechanism B-2.

Figure 7 shows the main chemical bonds, the profiles of k_{obs} variations as

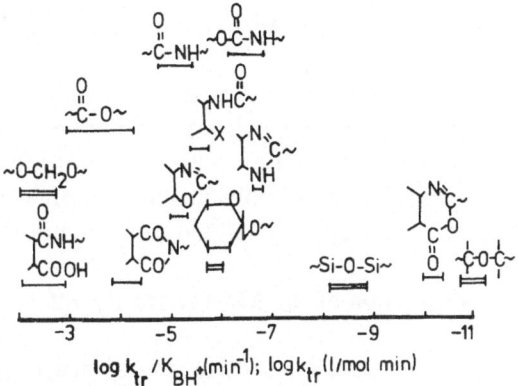

FIG. 8.   Kinetic parameters for the decomposition of simple model compounds with various chemically unstable bonds in aqueous solutions of acids. For mechanism A-1 ($\vdash\!\dashv$) $\log k_{tr}/K_{BH^+}$ is given; for mechanism A-2 ($\vdash\!\!\!=\!\!\!\dashv$) $\log k_{tr}$ ($mol^{-1} min^{-1}$) is given.[1]

a function of the concentration of the aggressive mediums and typical activation energies. Figures 8 and 9 represent regions in which the kinetic parameters occur for decomposition of simple model compounds with chemically unstable bonds in aqueous acid (mechanisms A-1 and A-2) and basic (mechanism B-2) media.

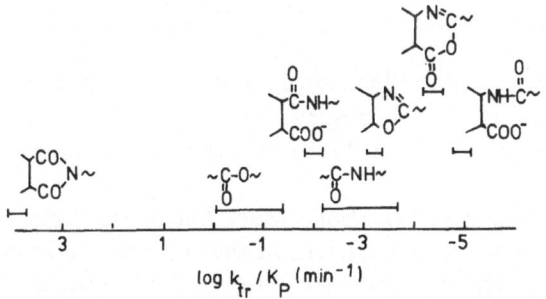

FIG. 9.   Kinetic parameters for the decomposition of simple model compounds with various chemically unstable bonds in aqueous solutions of alkalis (for mechanism B-2).

3.   PECULIARITIES OF THE CHEMICAL DEGRADATION PROCESSES IN POLYMERS: MAIN TYPES OF DECOMPOSITION IN POLYMERIC MOLECULES

Can the main principles, elaborated in the field of physical organic chemistry for low-molecular compounds, be applied to the chemical degradation of polymers? We shall try to answer this question in this section.

Paul Flory[34] was the first to enunciate the main principles of the reactivity of polymer functional groups being independent of the molecular weight for various reaction types. It is not infrequently found nowadays that the reactivity of various groups in polymers is abnormal; this is accounted for by the specific influence of the polymeric state.[35,36]

3.1. Quantitative Evaluation of the Macroviscosity Effect on Degradation Kinetics

Polymeric solutions can have a viscosity several orders higher than the solutions of low-molecular compounds. The macroviscosity of the reaction medium determines the diffusion rate of dissolved substances and, therefore, affects the rate of chemical reactions. The effect of diffusional factors on the kinetics of homogeneous chemical reactions has been dealt with in detail elsewhere.[37–39]

Even in monomolecular reactions the viscosity of the medium can influence (decrease) the rate due to the cage effect. The simplest way to describe the effect of diffusion on the kinetics of bimolecular reactions is by means of the following equation,[1,39]

$$k_{obs} = \frac{4\pi \delta D k_{tr} \beta}{k_{tr} + 4\pi \delta \beta D}$$

where k_{obs} and k_{tr} are the effective and true rate constants, respectively, δ is the sum of radii of the reacting particles, β is the proportionality factor and D denotes the sum of the diffusion coefficients of the reacting particles.

3.2. Specific Features of the Solid-phase Polymeric State

The conception of the peculiarities of the activated-complex structure in polymer reactions, that has been developed recently,[40,41] is extremely fruitful from the point of view of evaluating the reaction rates in a solid polymeric matrix. Due to the difficulty of molecular rearrangement in polymers the most favourable activated complex is not realised during the reaction; as a result, the process follows a higher profile of the potential energy surface. In this regard it is apparent that many reactions proceed at a lower rate in polymers than similar reactions in the liquid phase.[39–41] When passing from low-molecular (liquid phase) to macromolecular (solid phase) compounds, the rate constant may go down by 1, 3 or even 6 orders, merely due to the impossibility of realising the optimal structure of the activated complex. This effect is at its greatest at low temperatures when the mobility of polymeric chains is restricted. With increasing temperature, the rate of structural rearrangement in the polymer increases, and the configuration of the activated complex approximates to the optimal. The intensity of molecular motions in solid polymers determines not only the rate of rearrangement in polymers, but also the rate of diffusion of deleterious substances in the polymeric matrix. The connection between the activation energy E and the intensity of molecular motions in the polymeric matrix should involve the dependence of E on the temperature and the physical structure of the polymer. This is observed experimentally.[39-41]

The dependence of E on T yields abnormally high values for the Arrhenius parameters, i.e. the pre-exponential factor A and the activation energy E. The dependence of E on structure results in the experimentally observed 'step-wise' kinetics of reactions in a solid polymeric matrix.*

3.3. Effect of Functional Group Interaction on Reactivity

Functional groups are often located close to each other (10 Å). It is this fact that accounts for the abnormal reactivity of 'polymeric' functional groups. The abnormal reactivities of functional groups in polymers are classified as follows:

1. *Effect of the neighbouring group.* The order of this effect can be either long-range (the chain effect) or short-range (the influence of groups located next to the given functional group).
2. *Conformational effects.* These are the effects due to a change in the reactivity as a result of a change in the macromolecular conformation in the given reaction medium or in the course of the reaction. Such an influence is explained on the basis of the polar, resonance and steric effects, as well as on the basis of the formation of donor–acceptor complexes (hydrogen bonds, etc.).[39]

3.4. Main Types of Decomposition of Polymeric Molecules

Chemical bonds may be equally reactive; however, the end bonds in a molecule often exhibit increased reactivity which results in a complicated process of chemical degradation. In addition, polymers may contain other bonds with increased reactivity, the so-called 'weak bonds'.

---

* A solid polymer contains a certain set of molecules with different degrees of mobility which depend on the degree of molecular organisation. In its turn mobility determines reactivity. Thus we deal with a set of similar molecules that, at the same time, differ in reactivity by orders of magnitude.

3.4.1. Depolymerisation by the End-group Law

The fundamental theory behind this type of decomposition has been considered previously.[42]

Depolymerisation by the end-group law seldom occurs independently in the process of chemical degradation. Usually decomposition of this type takes place along with polymeric chain decomposition by the random law. Depolymerisation by the end-group law is exemplified by the degradation of polyoxymethylene with terminal hydroxyl groups (POM.OH) in aqueous solutions of bases.[43]

$$\sim\!\!CH_2OCH_2OCH_2OH + OH^- \longrightarrow \sim\!\!CH_2OCH_2OCH_2O^- + H_2O$$

$$\sim\!\!CH_2OCH_2O\cdots CH_2O^- \longrightarrow \sim\!\!CH_2OCH_2OH + HOCH_2O^-$$
$$H\cdots\ddot{O}\!\!-\!\!H$$

3.4.2. Decomposition by the Random Law

Chemical degradation[2−4,11,39] occurs most frequently by the random law.

3.4.3. Mixed Type of Decomposition

Macromolecular decomposition by the random law is often accompanied by depolymerisation of the fragments formed.[42] Such decomposition is part of the acid-catalytic degradation of polyoxymethylene with methoxyl end groups (POM.OCH$_3$), whose presence rules out depolymerisation of the initial compounds.[43−45]

The change in the mass of such a polymer is described by the following equation:[44,45]

$$m = m_0 \exp\{(2k_{dep}/k_r + 1)[\ln(2 - \exp - k_r t) - k_r t]\}$$

where: k_{dep} and k_r are the rate constants of decomposition by the depolymerisation mechanism and the random law, respectively; m_0 is the initial mass of PCM.OCH$_3$.

4. THE DYNAMICS AND MECHANISM OF DIFFUSION OF AGGRESSIVE MEDIA INTO POLYMERS

Solutions of electrolytes (acids, salts, bases) occupy a special place compared with other diffusants.[46−49] In the polymeric matrix they can dissociate into ions which move under the influence of two forces: the gradient of the chemical potential of the given ion type and the electric field generated by the motion of charged ions. If the polymers contain ionogenic groups, strong interaction of these groups with the diffusing ions may take place, resulting in diffusion (assuming a complex character).[50] Polyamides which are systems of aqueous electrolyte solutions are an example of such an interaction.[51]

4.1. Hydrophilic Polymers

Polymers that dissolve in water (polyvinyl alcohol, cellulose, etc.) and polymers with a restricted capacity for dissolving in water (polyamides, some polyesters, etc.) are classed as hydrophilic polymers.

An important problem (with regard to predicting polymer stability) is that of establishing the state of the electrolytes in the polymer,[49,50] for example, the degree of association, mobility of the ions, degree of hydration, etc. For polymers which dissolve well in water the parameters are the same as for the electrolyte solution.[51-54]

Many electrolytes in polymers obey the free-volume theory[49,50] which says that the relative variation in the self-diffusion coefficient of a low-molecular compound with a change in the parameters of the polymer–solvent system is described by the equation:

$$\log \frac{D^*}{D^*(o)} = \frac{B}{2 \cdot 3} \cdot \left(\frac{1}{f(o)} - \frac{1}{f} \right)$$

where B is the constant of the polymer–solvent system and f is the free-volume fraction of the system. The symbol (o) denotes the standard state system which is either the polymer or the solvent.

The most characteristic feature of electrolyte diffusion in hydrophilic polymers is the relationship between concentration and the diffusion coefficient which is usually affected by the following factors:[49-54]

a change in the degree of dissociation of the electrolyte;
the non-ideal behaviour of the polymer–electrolyte system;
swelling of the polymer leading to an increase in the mobility of the polymer;
binding of the electrolyte by ionogenic groups of the polymer.

Figure 10 shows the relationship between concentration and the diffusion coefficients of acids in polyvinyl alcohol. The dependences obtained are governed by the following: $D_{H_2SO_4} > D_{HSO_4^-} > D_{SO_4^=}$, while $D_{H_3PO_4} > D_{H_2PO_4^-} > D_{HPO_4^=} > D_{PO_4^\equiv}$.

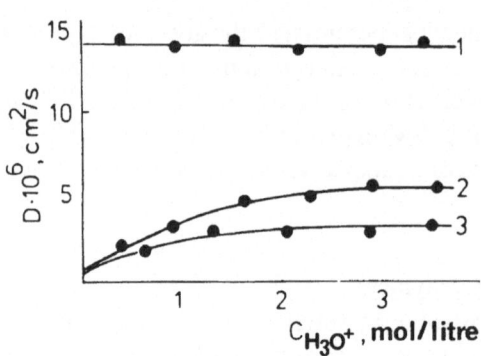

FIG. 10. Dependence of the integral coefficients of diffusion of acids in polyvinylalcohol films at 25 °C on the concentration[55] of (1) HCl, (2) H_2SO_4, (3) H_3PO_4.

4.2. Hydrophobic Polymers

This group includes polyolefins, polyesters, polyacetals and polysiloxanes. Investigations of the diffusion of electrolytes in hydrophobic polymers are associated to a considerable extent with the use of the polymers as protective anticorrosive coatings.[56] Hydrophobic polymers are practically devoid of polar groups which accounts for the low solubility of electrolyte solutions in them.[49] The electrolytes are not ionised in this kind of polymeric matrix.

Calculations indicate[57] that the electrical conductivity of the polymer would be 10 orders higher than that determined experimentally, if the whole amount of HCl dissolved by polyethylene from concentrated hydrochloric acid solutions was ionised. The data obtained in absence of ionisation of the electrolytes are represented in Fig. 11.

Thus, for the diffusion of electrolyte solutions in hydrophilic polymers the thermodynamic (e.g. k_{dis}) and diffusion (e.g. D^*) parameters are close to the respective parameters in solutions. A decrease in the mobility of the electrolyte molecules and ions in polymers, compared to their mobility in solutions, is due to the increased length of the diffusion path, as well as the interaction of the diffusing particles with the polymer. In hydrophilic polymers which dissolve in water to a limited degree, the dissolved water does not form a continuous water phase in the polymeric matrix, and the ions diffuse by way of activated leaps between the polar groups; as a result a considerable decrease in D for the electrolyte, as compared to its value in solution, is observed.

The mechanism of diffusion of electrolytes in hydrophobic polymers is similar to the transfer of gases and vapours. Therefore, for electrolytes with

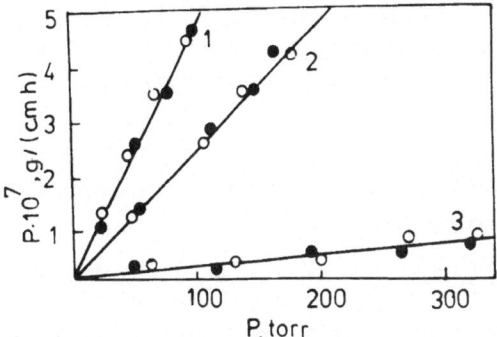

FIG. 11.   HCl permeability through polymeric films from aqueous solution (●) and from the gas phase (dry HCl) (○) as a function of HCl pressure:[67] (1) polyethylene, (2) polypropylene, (3) polyethylene terephthalate.

a high vapour pressure, the values of C^0_{electr} and D^0_{electr} are close to those obtained for water in these polymers. The electrolytes with a low vapour pressure show extremely low values of C^0_{electr} and D^0_{electr}. The sorption and diffusion characteristics typical of electrolytes in hydrophilic and hydrophobic polymers are summarised in Table 2.

TABLE 2

TYPICAL SORPTION AND DIFFUSION CHARACTERISTICS OF AQUEOUS SOLUTIONS OF ELECTROLYTES IN HYDROPHILIC AND HYDROPHOBIC POLYMERS

Typical sorption or diffusion characteristics	*Polymers*	
	Hydrophilic	*Hydrophobic*
Concentration of absorbed electrolyte	High	Low
Henry's law	May not be kept	Is kept
Dissociation into ions	Close to dissociation in water	Practically absent
Values for diffusion coefficient	Close to diffusion coefficient in water	Low for non-volatile and high for volatile electrolytes
Main factor determining diffusion rate for electrolytes	Size of ions and dissociation of electrolytes	Partial pressure of electrolyte over solution
Effect of electrolyte sorption on electrical resistance of polymer	Resistance decreases by several orders	Practically no effect

5. A MATHEMATICAL DESCRIPTION OF DEGRADATION PROCESSES; PREDICTION OF THE STABILITY OF POLYMERIC PRODUCTS

5.1. Polyamides

Within the temperature range 20–200 °C[58] polycaproamide degradation in water is described by the equation:

$$k^*_{H_2O}(\text{min}^{-1}) = 1 \cdot 6 \times 10^7 \exp(-23{,}000/(RT))$$

The high reactivity of quite a small proportion of the amide bonds ($\simeq 1\%$) is accounted for by mechanical activation due to the residual tensile stresses at individual chains arising from elongation with subsequent fast cooling.[2,39,40] A comparison between the decomposition rate constant for the normal part of an amide (k_{H_2O}) and that for the reactive bonds ($k^*_{H_2O}$) indicates that the mechanically activated amide bonds in polycaproamide are more reactive by more than three orders than other amido bonds.

The change in k_{obs} (min^{-1}) in aqueous solutions of sulphuric acid is described by the following equation:[59]

$$k_{obs} = 2 \cdot 5 \times 10^{-9} \exp(-23{,}000/(RT)) \frac{H^+_s}{1 + (h_A/5 \cdot 0)}$$

With an increasing degree of orientation of polycaproamide fibre, the degradation rate decreases as a result of the changed accessibility [60,61] of amido bonds to the acid or water.

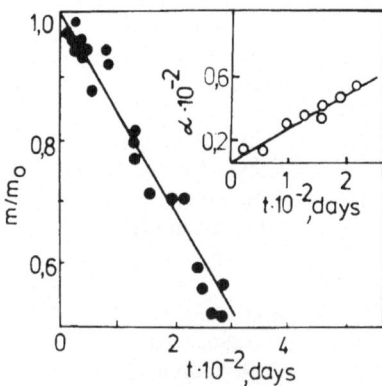

FIG. 12. Variations in the relative mass of fibres (m/m_0) and in the transformation degree (α) of amino bonds in polycaproamide as a function of the time of implantation in the subcutaneous fat of rabbits.[50]

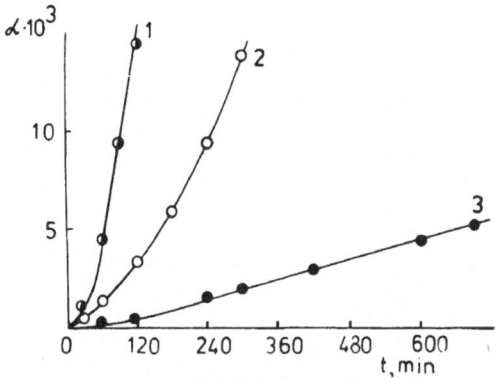

FIG. 13.   Dependence of the degree of transformation of ester bonds on the time of degradation of PETPH films of various thicknesses in a 53 % solution of sulphuric acid at 116 °C.[62] (1) 5 μm, (2) 20 μm, (3) 80 μm.

As a rule the degradation of amides takes place in the internal kinetic area (Fig. 12), that is, where there is no diffusion control of the reaction and there is enough reactant inside the polymer.

5.2. Polyesters

The chemical degradation of polyethylene terephthalate (PETPH) has been thoroughly studied.[62] For PETPH it is important to know the diffusion dynamics of aggressive media. The following equation has been derived for D_{H_2O}:[62]

$$D_{H_2O} = 4{\cdot}5 \times 10^{-2} \exp(-10{,}000/(RT))$$

Figure 13 represents a typical dependence of the transformation of ester bonds as a function of time for the degradation of PETPH. The degradation takes place in a reaction zone which varies with time. In the general case the degradation process which occurs at the surface of the sample cannot be disregarded, since the ester bonds that are on the surface come into contact with a greater number of acid molecules than the ester bonds inside PETPH. It has been found from the equation

$$C_n = k_{obs} C_{cat}^o t$$

where C_n is the number of reactive bonds, that

$$k_{obs} C_{H_2SO_4}^o = 1 \times 10^{-8}\, \text{s}^{-1}\, \text{equiv cm}^{-3}$$

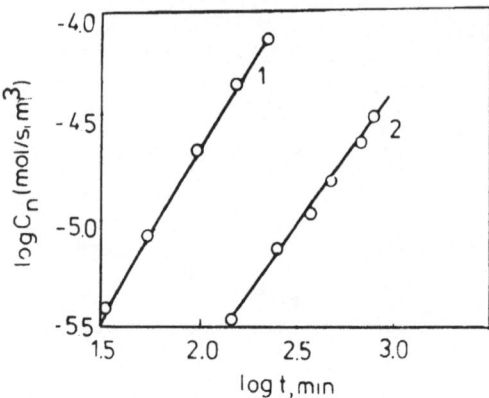

FIG. 14.   Dependence of $\log C_n$ on $\log t$ for the degradation of PETPH film.[62] (1) 20 μm, (2) 80 μm.

hence, using the following equation[1,49,50]

$$C_n = \frac{4}{\pi^{1/2}} k_{\text{obs}} C_{\text{cat}}^{\text{o}} D_{\text{cat}}^{1/2} l^{-1} t^{3/2}$$

where l is the thickness of the sample. $D_{\text{H}_2\text{SO}_4}$ can be found (Fig. 14).

PETPH films of various thicknesses yield different values for the diffusion coefficients which is due to the varying structure of the films.

The overall process of degradation of PETPH films in aqueous solutions of sulphuric acid is described by the equation:

$$C_n = k_{\text{obs}} \cdot C_{\text{H}_2\text{SO}_4}^{\text{o}} \cdot t \left\{ 1 - \frac{8}{\pi^2} \right.$$

$$\times \sum_{m=1}^{\infty} \left[1 - \exp(-(2m-1)^2) \frac{\pi^2 t}{l^2} \, 2{\cdot}9 \times 10^{-7} \exp(-12{,}000/(RT)) \right]$$

$$\times \left. \frac{4l^2}{(2m-1)^4 \pi^2 t \times 2{\cdot}9 \times 10^{-7} \exp(-12{,}000/(RT))} \right\}$$

The following facts hold for PETPH degradation in alkaline solutions:[62]

1.   The kinetic order of the reaction with respect to the polymer is zero provided the reaction products are removed.
2.   The molecular weight of PETPH is constant up to the end of the process.

3.   No absorption bands due to carboxyl groups, resulting from decomposition of the ester bond, are present in PETPH films.

The following equations describe the change in the mass of PETPH films and fibres during degradation in aqueous solutions of alkalis:

$$m_n = m_n^\circ - \frac{6\cdot4 \times 10^6 \exp(-16{,}500/(RT))a_{H_2O}st}{1 + \dfrac{4 \times 10^2}{b_0}}$$

$$m_n = m_n^\circ \left[1 - \frac{6\cdot4 \times 10^6 \exp(-16{,}500/(RT))a_{H_2O}t}{\left(1 + \dfrac{4 \times 10^2}{b_0}\right)\bar{r}_0\rho}\right]^2$$

where m_n is in g, t is in min, s is in cm^2, ρ in g cm^{-3} and $\bar{r}_0$ in cm.

In connection with the problem of working out materials that are self-resolving in a living body, a detailed investigation has been carried out into the degradation of highly-crystalline polyglycolide.[2,3,61] Polymeric filaments of polyglycolide (a self-resolving polymer) consist of fibrils that contain crystallites separated from each other by amorphous sections. The variation in the polymer mass is generally expressed as:

$$m/m_0 = 1 - 0\cdot2[1 - \exp(-k_r t)\bar{c}_{H_2O}^v] - 0\cdot8\left[\frac{2k_{obs}^{s-v} t\bar{c}_{H_2O}^{s-v}}{N\bar{b}_0\rho} - \left(\frac{k_{obs}^{s-v} t\bar{c}_{H_2O}^{s-v}}{N.\bar{b}_0\rho}\right)^2\right]$$

$$- \frac{2k_{tr}^s tc_{OH^-}}{K_{eq}\bar{r}_0\rho} + \left(\frac{k_{tr}^s tc_{OH^-}}{K_{eq}\bar{r}_0\rho}\right)^2$$

where K_{eq} is the equilibrium constant of formation of active species, s stands for 'surface' (the surface concentration), and v stands for the 'volume' (the bulk concentration).* In aqueous solutions of alkalis the degradation of many types of polyarylates proceeds by the same mechanism as that of PETPH. The variation in the polymeric film mass is described by the equation:

$$m = m_0 - 6 \times 10^3 \exp(-13{,}500/(RT))a_{H_2O}b_0st$$

---

* Degradation of the amorphous phase (20%) occurs in the bulk; degradation of the crystallites occurs on the surface of the polymer.

5.3. Cellulose and its Derivatives

Let us consider the degradation of cellulose and its derivatives under an homogeneous condition in acid solutions.[63,64] The difference between the reactivities of the ultimate and penultimate glycoside bonds in the molecules is small (1·5 fold). The degradation of cellulose and its derivatives proceeds by the same mechanism as has been established for oligomers (see Section 1).

In the amorphous phase the degradation proceeds, as a rule, in the internal kinetic area by the random law. The rate constants can be calculated from the equation:

$$\frac{2}{\bar{P}_w} - \frac{2}{\bar{P}_{w_0}} = kNt$$

where N denotes the proportion of glycoside in the amorphous portion.

5.4. Polyacetals

The main decomposition products[65] of polyacetals are formaldehyde and cyclic polyacetals. The rate of the mass decrease is described by the equation:

$$\frac{\Delta m}{\Delta t} = 1/c_n \cdot v \left\{ k_{surf}[H_s^+]_{surf} \frac{1}{zl} + k_{vol}[H_s^+]_{vol} \right\}$$

where l is the average depth of the reaction zone in the polymeric article and z is the number of monolayers.

5.5. Polycarbonates

Polycarbonate (diflon) degradation[50] has been studied in detail in acid and in basic media. The degradation process takes place both from the surface of the polymeric product (acid specific catalysis) and inside it (catalysis by the water molecules). In the former case the degradation proceeds within the external diffusive–kinetic area, and the overall variation in the polymer mass is described by the equation:

$$m_n = m_n^o + \Delta m_{H_2O}^\infty [1 - 8/\pi^2 \exp(-\pi^2 \bar{D}t/(l^2))] - k^s h_0 st$$

where m_{H_2O} stands for the quantity of water in the polymer, corresponding to the equilibrium value of sorption and k^s is the rate constant of degradation of the polymeric product from the surface, independent of catalyst concentration.

In the latter case the degree of transformation of carbonate bonds (degradation by the random law) is satisfactorily described by the equation:

$$c_n = k_{obs} c_{cat}^o t$$

5.6. Polysiloxanes

The kinetics and mechanism of degradation of polysiloxanes have been reported in detail elsewhere.[11,50,66,67] It was shown that for polydimethyl-siloxane rubber[1] (degradation in acid–water solution) the rate constant of reaction is related to some parameter (which is proportional to the degree of degradation) by the equation

$$\ln\left[1 - \left(\frac{\alpha_0 - \alpha_t}{\alpha_0}\right)^{1/2}\right] = -kt$$

where 0 and t refer to times 0 and t respectively.

5.7. Poly(vinyl chloride)

The kinetics and mechanism of degradation of poly(vinyl chloride) have been considered.[68] It has been shown that the structure of the PVC macromolecule is responsible for the low stability of PVC.

The presence in the PVC macromolecule of a number of β-chloroallyl, oxygen-containing chloroallyl and other groups is very important. The appropriate experimental evidence has been discussed[68] and the new evidence has been used to construct well-grounded PVC degradation and stabilisation schemes on the basis of soundly based kinetic equations.

6. EFFECT OF MEDIUM ON THE MECHANICAL PROPERTIES OF POLYMERS

The use of many polymeric articles and materials involves the simultaneous influence of liquid media and mechanical stresses, e.g. in the implantation of polymers into the living body, in the use of polymers as liquid-media filters, in the production of pipelines, in polymeric parts for electric pumps, in polymeric containers for the transportation of liquids, etc.

The mechanical behaviour of such systems is, naturally, a most pressing problem from the practical point of view.[69-73] Polymers are distorted by the load applied. Creep, i.e. the property of solids to accumulate deformations under the action of constant stresses, is an important parameter.

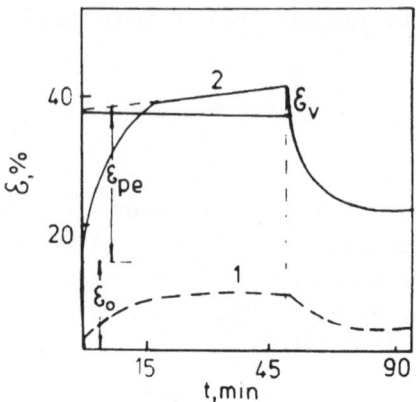

FIG. 15.   Curves for polyarylate creep (1) in air and (2) in water.[69] (ε_0) elastic, (ε_{pe}) plasto-elastic, (ε_v) viscous; $T = 20\,°C$, $\sigma = 8\,MPa$.

The creep of glassy polymers in air is usually small and amounts to a few per cent. However, if, for instance, an industrial polymer—polyarylate (F-2)—is subjected to uniaxial tension in water, the creep amounts to dozens of percent (Fig. 15), although the solubility of water in the polymer does not exceed 3 % by weight.[69−71]

The overall deformation in creep is a combination of elastic, plasto-elastic and viscous (set-in) deformations (Fig. 15). The elastic deformation develops almost immediately and is measured by Hooke's law

$$\varepsilon_0 = \sigma(t)/E$$

where E is Young's modulus. The viscous deformation is small as a rule, and is found through Newton's law

$$\varepsilon_v = \frac{1}{\eta}\sigma(t)\,dt$$

where t is the time, $\sigma(t)$ denotes the stress at time t (it is to be noted that with the increasing time of loading the viscous deformation for many of the polymers investigated will not exceed 2 % up to the moment of breakage).

The plasto-elastic deformation, which has the greatest value, develops with time. However, no equation has been derived so far to describe this type of deformation versus time, stress or temperature in liquid media. Semiempirical equations[71] are commonly used, either the Boltzmann–Voltaire equation (the heredity theory) or the Alfrey equation (the theory of retardation time set).

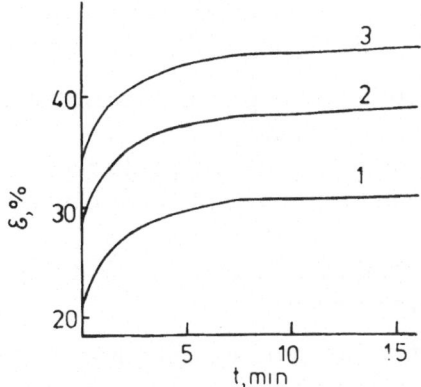

FIG. 16.   Curves for PETPH creep in water at various temperatures. (1) 20 °C, (2) 40 °C, (3) 60 °C; $\sigma = 20$ MPa.

However, both equations contain unknown functions which are usually found by trial and error. Besides, these approaches have been elaborated for solid-state creep in air and do not take into account the specific features of polymer creep in liquid media.

If the diffusion rate of the liquid medium is greater than the polymer creep in air the dependence of plasto-elastic deformation on time is found to be:[69-71]

$$\varepsilon(t) = \varepsilon_x [1 - \exp(-t/\theta)] \qquad (15)$$

where θ is the retardation time, i.e. in this case one creep mechanism, associated with the environmental influence, prevails. The creep mechanism in the liquid medium differs from that in air. In the liquid medium the creep occurs due to the motion and rearrangement of structural elements of the same type, and there exists only one retardation time, while in air a set of retardation times is observed.

Figure 16 contains typical data obtained for the creep of PETPH films in water and values of θ, obtained using eqn (15), are as follows.

T (°C)	20	40	60
θ (min)	1·8	1·6	1·6

It is noticeable that retardation times show only a slight variation with increasing stress and temperature, while the same factors greatly affect the equilibrium deformation.

Under mechanical stress the polymer creep in liquid media changes, as compared to the creep in air, due to the following processes:[69-72]

1. Water adsorption leading to a decreased surface energy at the polymer–medium boundary.
2. Water sorption leading to a change (decrease) in the intermolecular interaction and to an increased distance between macromolecular fragments.
3. Chemical degradation leading to the disruption of chemically unstable stressed bonds.

The role of the first two processes (adsorption and sorption) can be revealed in the following way.

Using KNO_2 solutions of various concentrations, the surface energy at the polymer–medium boundary (determined by the wetting angle) is practically the same; with increasing concentrations of KNO_2, however, the activity of water in the solution decreases, thus making it possible to vary over a broad range the quantity of water dissolved in F-2 polyarylate and polyethylene terephthalate. Figure 17 shows the equilibrium deformation decreases with increasing concentrations of KNO_2. At the same time, in dilute solutions of the surface-active substance, sodium dodecyl sulphate (when the water concentration in the polymer is practically the same), ε_x did not change. Thus sorption of the medium has a decisive effect on creep in this case.

Creep involves changes both at the supermolecular and molecular levels. X-ray diffraction at small angles has demonstrated that for a partially

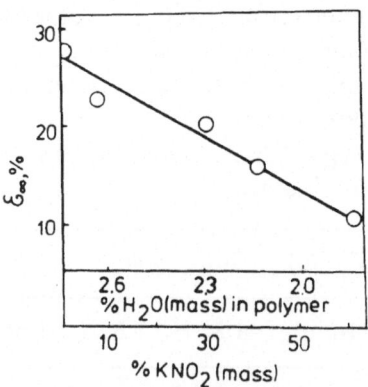

FIG. 17.   Dependence of the equilibrium deformation (ε_x) of F-2 polyarylate on KNO_2 concentration and on the quantity of water in the polymer;[70] 20 °C, $\sigma = 80$ MPa.

crystalline polymer (polyethylene terephthalate, polycaproamide) there exists a linear dependence (Fig. 18) of the macrodeformation on the long period value (with the degree of crystallinity retained); i.e., with stresses below the breaking value and under the influence of the medium, the deformation of macromolecular fragments in amorphous regions occurs. Simultaneously, plasto-elastic deformation, i.e. chain bending with the atomic spacings retained, is observed.

In amorphous regions, macromolecular conformational transformations, mostly *gauche–trans* transitions, take place. Figure 18 shows the variation in the relative intensity of absorption in infrared spectra of the bonds of *trans* and *gauche* isomers. Observations show that the increasing deformation is accompanied by macromolecular straightening, with the number of rotational isomers in coiled conformations decreasing.

It is suggested[69–72] that, with conformational transformations requiring certain free space, the greater the free volume of the polymer–medium system, the greater ε_α, i.e.

$$\varepsilon_\alpha = A \exp(Bf) \tag{16}$$

where f is the free-volume fraction of the polymer–medium system and A and B are parameters that are typical of the polymer–medium system but are independent of the temperature and concentration of the medium in the polymer.

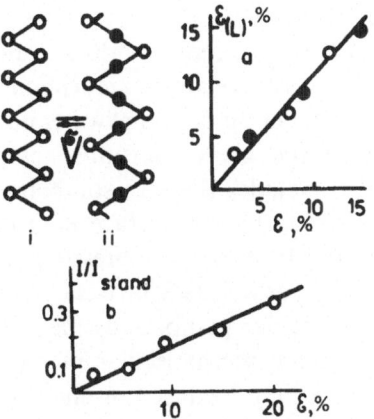

FIG. 18.   (a) Dependence of the long period value[70] on macrodeformation of samples of polyethylene terephthalate (●) and polycaproamide (○); 20 °C. (b) Variation in the relative intensity of infrared absorption spectra for the bonds of *trans* (i) and *gauche* (ii) conformers with ε (%);[70] polyethylene terephthalate, 20 °C, $\sigma = 100$ MPa, 845 cm^{-1} (δ C—H)—.

Equation (16) implies that equilibrium deformation ε_{χ} is not directly connected with temperature and mechanical stress, but is mainly dependent on the free-volume fraction whose value determines the possibility of conformational transformations (decreases the force of intermolecular interactions and increases free space for the transformations).

The influences of temperature, mechanical stress and medium manifest themselves in the free volume variations since they are a function of these parameters.

Let us choose the equilibrium deformation of a polymer in air ε_{χ}^{0} as the standard state. Hence

$$\varepsilon_{\chi} = \varepsilon_{\chi}^{0} \exp\left[B(f - f^{\circ})\right] \tag{17}$$

where f° is a standard state. The free volume fraction is

$$f = f_{p}\psi_{p} + f_{m}\psi_{m} \tag{18}$$

where f_{p} and f_{m} are the free-volume fractions of the polymer and medium, respectively, and ψ_{p} and ψ_{m} denote volume fractions.

Introducing $f_{m} - f_{p} = \beta$ and with $f^{\circ} = f_{p}$, we obtain

$$\log \varepsilon_{\chi} = \log \varepsilon_{\chi}^{0} + B\beta\psi_{m} \tag{19}$$

Equation (19) makes it possible to predict variations in the ultimate high-elastic deformation as a function of the volume fraction of the medium introduced into the polymer.[70]

Under tensile stress $f^{\circ} = f_{p}$ increases with the value of σ if the polymer does not crystallise. F-2 polyarylate is amorphous; special experiments have demonstrated that under the action of water no crystallisation occurs.

For polymers below the glass transition temperature (T_{g}) double sorption takes place; it includes the true solubility of the low-molecular compound and its sorption by the surface micropores. The volume of micropores can be found from the compression isotherms, the sorption isotherms of inert vapours, as well as from data on mercurial porometry.[71]

Truly soluble low-molecular compounds may be expected to exercise much greater influence on the intermolecular interaction and, consequently, on the creep than the medium localised in micropores.

The volume fraction of the medium activating the creep process will amount to

$$\psi_{m} = \psi_{m}^{0}(1 + n\sigma) - \psi_{pore} \tag{20}$$

and

$$\log \varepsilon_{\chi} = \log \varepsilon_{\chi}^{0} + B\beta[\psi_{m}^{0}(1 + n\sigma) - \psi_{pore}] \tag{21}$$

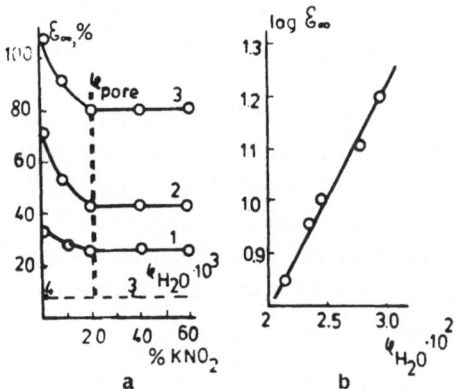

FIG. 19. (a) Dependence of the equilibrium deformation (ε_x) of polycarbonate on the quantity of water (ψ_{H_2O}) in the polymer (concentration of KNO_2): $\sigma = 48\,MPa$; (1) 20 °C, (2) 60 °C, (3) 90 °C. (b) Dependence of the equilibrium deformation (ε_x) of F-2 polyarylate on the quantity of water in the polymer (ψ_{H_2O});[70] $\sigma = 80\,MPa$, 20 °C, $\beta = 40 \pm 2$, $n = (2\cdot5 \pm 0\cdot5) \times 10^{-3}$.

Equation (21) permits one to predict the creep process. Thus:

1. In inert liquid media the polymer creep is the same as in air, since the medium is localised at micropores (polyethylene, polypropylene in water).

2. In moderately inert media double sorption takes place, and the creep changes in a complicated way depending on the concentration of the medium in the polymer.

For example (Fig. 19), for polycarbonate creep in aqueous solutions of KNO_2, in concentrated solutions of the salt, the whole water in polymer is practically localised in micropores (the overall volume has been established by the mercurial porometry method). In more dilute salt solutions the creep increases, since some part of the water dissolves in the polymer and activates the creep process.

For F-2 polyarylate in all KNO_2 solutions the creep increases, since $\psi_{H_2O} > \psi_{pore}$ for all $\sigma > 30\,MPa$. Below this value of σ, which is called critical, the observed results are the same as for polycarbonate in concentrated solutions of KNO_2.

Thus, the first step in predicting the polymer creep in liquid media has been made. The free volume of the polymer–medium system has been chosen as the determining parameter.[69–72]

7. HYDROLYTIC DEGRADATION AS A METHOD OF POLYMER MODIFICATION

Chemical degradation in polymers has a double significance. On the one hand, the chemical, physical and mechanical properties of polymers can change, when the latter are in contact with aggressive media, to such an extent that the articles become unfit for service. On the other hand, chemical degradation is used for modification, recovery, pickling and for other purposes, i.e. it is used (or may be used) to improve certain service properties of polymeric articles.

7.1. Elaboration of Optimal Conditions to Produce Rough Films from Polyethylene Terephthalate

The technology for producing rough-surfaced films for electronic computers and video- and sound-recording equipment—synthetic paper— has been intensely developed recently[50] and polystyrene, polyethylene and polyethylene terephthalate are used as the starting materials. One synthetic paper production method consists of the special chemical treatment (pickling) of polymeric film. To find optimal pickling conditions

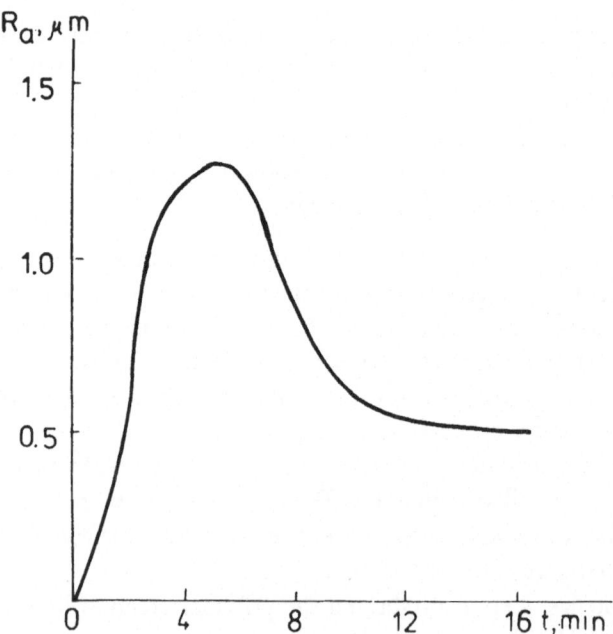

FIG. 20.   Dependence of R_a on the PETPH film pickling time (thickness 500 μm, 40 % crystallinity) in 49 % KOH solution at 108 °C.[74]

for PETPH films it is necessary to know the pickling mechanism and the dependence of the degree of roughness on time, temperature, thermo-dynamic parameters and polymer structure.[74,75]

As can be seen in Fig. 20, in the course of pickling a thick PETPH film, the roughness index R_a first increases, then decreases. It was suggested earlier that the formation of roughness is associated with pickling the amorphous portion of the polymer. This is not so (Fig. 21) however, at least not for PETPH films. In this case the formation of roughness is associated with revealing in the course of pickling microdefects that form during the production of the polymer from melts at a depth of 25–30 μm. An expression has been obtained[50] to show variations in the roughness parameter, R_a, versus time, temperature and the thermodynamic parameters of the medium:

$$R_a = \frac{6 \cdot 4 \times 10^6 \exp(-16,500/(RT)) a_{H_2O} st}{1 + \dfrac{4 \times 10^2}{b_0}}$$

where a_{H_2O} is a proportionality factor dependent on the nature of the polymeric material.

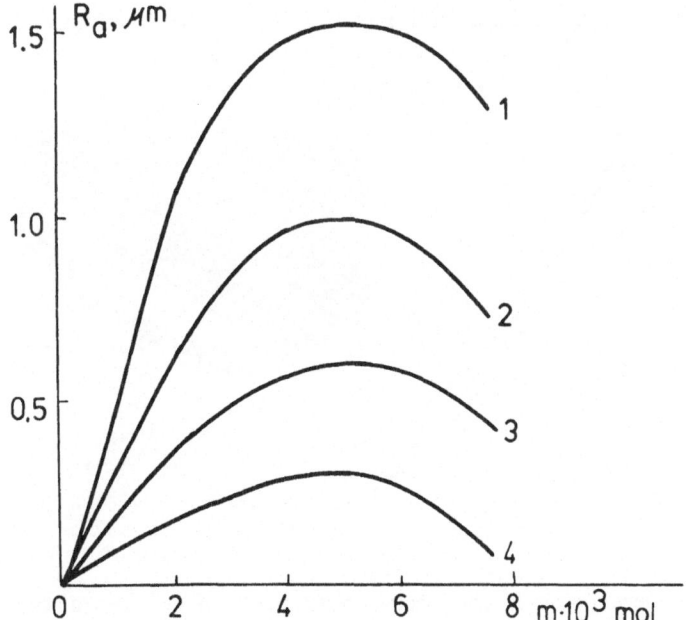

FIG. 21. Dependence of R_a on the mass of the removed layer for PETPH films of various degrees of crystallinity. (1) 0·5%, (2) 30%, (3) 40%, (4) 50%.[75]

7.2. Modification of the Surface of Cellulose Diacetate Fibres with Aqueous Alkali Solutions

A deficiency of synthetic fibres, including polyester, is their high electrification. The accumulation of static charges gives rise to technological and service difficulties. In acetate fibres treated with basic solutions a layer of recoverable cellulose forms as a result of ester bond hydrolysis. No hydrolysis of glycoside bonds is observed. The inside of the fibre is a synthetic polymer, while on the outside it is flax or cotton. The task resolves itself into finding the proper conditions, such that the hydrolysis rate is much higher than the diffusion rate, i.e. degradation should proceed within a narrow region near the surface (Fig. 22). The above conditions can be achieved in two ways:

1.   By the use of concentrated solutions of strong alkalis (an increased hydrolysis rate with constant D_{OH^-}).
2.   By the use of an alkali with a cation of large volume, for instance $(C_4H_9)_3NOH$ (decreased D_{OH^-}, with the reaction rate retained).

7.3. Production of Filters with Designed Size of Micropores

Nuclear filters are coming into use for ultrafiltration, sterilisation and separation of chemical and biological media.[1,50] The production of filters

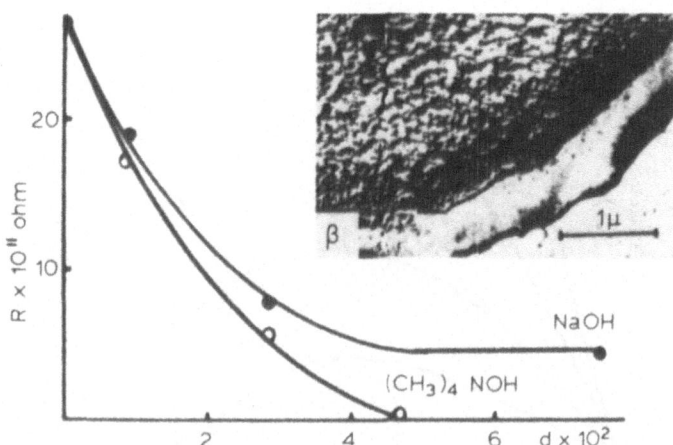

FIG. 22.   Dependence of the electric resistance[76] of cellulose diacetate cloth surface on degree of hydrolysis in different solutions. Microphoto: filament of cellulose diacetate after hydrolysis.

from polycarbonate (diflon) and polyethylene terephthalate films involves three steps:

1. Irradiation of the polymeric film with heavy ions which, as they pass, cause the formation of local regions of degraded polymer (tracks).
2. Exposure of the irradiated film to ultraviolet light for greater degradation of oligomeric polymeric molecules in the track zone.
3. Treatment of the irradiated film with an alkaline solution, resulting in the degradation of oligomeric and polymeric fragments in the track zone, as well as of the non-degraded polymer beyond the track zone, and also in the desorption of the decomposition products into the environment.

Generally, the increase in the mass of the decomposition products is a result of the degradation reactions proceeding:

1. On the surface of the polymeric film, resulting in a thin film;
2. Within the track zone;
3. On the side surface of the tracks, resulting in an increased diameter of the microchannels.

The change in the mean radius of the microchannels in the PETPH film on treatment with aqueous solutions of alkalis is described by the equation:

$$\bar{r} = \bar{r}_0 + 1 \cdot 6 \times 10^6 / \rho \exp(-16{,}500/(RT))b_0 a_{H_2O} t$$

8. CONCLUSION

The present article deals with the methods of investigation of hydrolytic degradation processes in polymers. It has been demonstrated that a study of the kinetics laws of hydrolysis gives a deeper insight into the mechanism of the process and makes it possible to derive equations describing the degradation process over a wide range of reaction conditions.

Material demonstrating the success of such investigations has been presented in a number of books and reports in the preparations of which the author has participated.[1-4,11,49,60,68,74]

Very little work has been done so far[77-80] on the investigation of the reduction of rates of hydrolytic decomposition or on the search for methods of polymer stabilisation against hydrolytic degradation processes.

REFERENCES

1. MOISEEV, YU. V. and ZAIKOV, G. E., *Khimicheskaya Stoikost' Polymerov v Agressivnykh Sredakh*, Moscow, Khimiya, 1979.
2. ALKSNIS, A. F., ZAIKOV, G. E. and KARLIVAN, V. P., *Khimicheskaya Stoikost' Poliefirov*, Riga, Zinante, 1978.
3. ZAIKOV, G. E. and PRIVALOVA, L. G., Problemy ispol'zovaniya polymerov v khirurgii, In: *Khimiya Nashimi Glazami*, Moscow, Nauka, 1981, pp. 469–87.
4. BOCHKOV, A. F. and ZAIKOV, G. E., *Chemistry of the O-Glycosidic Bond: Formation and Cleavage*, Oxford, Pergamon Press, 1979.
5. LACEY, R. N., *J. Chem. Soc.* (1960), 1633.
6. FRAENKEL, G. and FRANCONI, C. J., *J. Amer. Chem. Soc.*, **82** (1960), 4478.
7. O'CONNOR, C., *Quart. Rev.*, **24** (1970), 533.
8. MOISEEV, YU. V., BATYUKOV, G. I. and VINNIK, M. I., *Izvetiya AN SSSR, ser. fiz.*, **20** (1962), 1306.
9. LOCKERENTE, R. S. DE NAGY, O. B. and BRUYLANTS, A., *Org. Magn. Resen.*, **2** (1970), 179.
10. NECHAEV, P. P., MOISEEV, YU. V. and ZAIKOV, G. E., *Vysokomolek. Soed.*, **A14** (1972), 1048.
11. MOISEEV, YU. V., MARKIN, V. S. and ZAIKOV, G. E., *Uspekhi Khimii*, **45** (1976), 510.
12. VINNIK, M. I. and MEDVETSKAYA, I. M., *Zhurn. Fizicheskoy Khimii*, **41** (1967), 1775.
13. STIDILA, F. H., *J. Org. Chem.*, **37** (1972), 178.
14. VINNIK, M. I. and MOISEEV, YU. V., *Tetrahedron*, **19** (1963), 1441.
15. MOISEEV, YU. V., BAKHRAKH, E. YA. and VINNIK, M. I. *Zhurn. Fizicheskoy Khimii*, **37** (1963), 784.
16. ORLOV, I. G., MOISEEV, YU. V. and VINNIK, M. I., *Reaktsionnaya Sposobnost' Organicheskikh Soedineniy*, **2** (1965), 180.
17. NECHAEV, P. P., MOISEEV, YU. V. and ZAIKOV, G. E., *Intern. J. Chem. Kinetics*, **6** (1974), 245.
18. NECHAEV, P. P., MOISEEV, YU. V. and ZAIKOV, G. E., *Vysokomolek. Soed.*, **A15** (1973), 702.
19. KORSHAK, V. V., TSEITLIN, G. M. and PAVLOV, A. I., *Doklady AN SSSR*, **163** (1965), 116.
20. YAKUBOVICH, V. S., *Doklady AN SSSR*, **159** (1964), 630.
21. KULAGIN, V. N., TSEITLIN, G. M., MOISEEV, YU. V. and ZAIKOV, G. E., *Izvestiya AN SSSR, ser. khim.* (1974), 1770.
22. DONSKIKH, A. I., KULAGIN, V. N., MOISEEV, YU. V., TSEITLIN, G. M. and ZAIKOV, G. E., *Izvestiya AN SSSR, ser. khim.* (1978), 562.
23. SIYGUR, YU. R. and KHALDNA, YU. L., *Reaktsionnaya Sposobnost' Organicheskikh Soedineniy*, **7** (1970), 211, 412, 431.
24. VINNIK, M. I. and LIBROVICH, N. B., *Reaktsionnaya Sposobnost' Organicheskikh Soedineniy*, **7** (1970), 1221.
25. KHALDNA, YU. L. and RODIMA, T. K., *Reaktsionnaya Sposobnost' Organicheskikh Soedineniy*, **10** (1973), 719.

26. VINNIK, M. I. and LIBROVICH, N. B., *Izvestiya AN SSSR, ser. khim.* (1975), 2211.
27. MATHIEU, J. and ALME, A., *Printsipy Organicheskogo Sinteza*, Moscow, Mir, 1964.
28. EDWARD, J., *Chem. Ind. (London)*, (1955), 1102.
29. ARMSTRONG, H. E. and GLOVER, W. H., *Proc. Roy. Soc.*, **80** (1968), 312.
30. KHALTURINSKII, N. A., MOISEEV, YU. V. and ZAIKOV, G. E., *Doklady AN SSSR*, **198** (1971), 149.
31. ANDRIANOV, K. A. and YAKUSHINA, S. E., *Vysokomolek. Soed.*, **7** (1965), 613.
32. RAZUMOVSKII, L. P., MOISEEV, YU. V. and ZAIKOV, G. E., *Izvestiya AN SSSR, ser. khim.* (1973), 2448.
33. RAZUMOVSKII, L. P., MOISEEV, YU. V. and ZAIKOV, G. E., *Doklady AN SSSR, ser. khim.* (1975), 1502.
34. FLORY, P. J., *Principles of Polymer Chemistry*, Ithaca, Cornell University Press, 1953.
35. PLATE, N. A., *Kinetika i Mekhanism Obrazovaniya i Prevrascheniya Makromolekul*, Moscow, Nauka, 1958, p. 250.
36. TUTORSKII, I. A., NOVIKOV, S. A. and DOGADKIN, B. A., *Uspekhi Khimii*, **35** (1966)), 191.
37. ENTELIS, S. G. and TIGER, R. P., *Kinetika Reaktsii v Zhidkoy Faze*, Moscow, Khimiya, 1972.
38. EMANUEL, M. N., ZAIKOV, G. E. and MAIZUS, Z. K., *Rol' Sredy v Radikal'notsepnykh Reaktsiyakh Okisleniya Organicheskikh Soedinenii*, Moscow, Nauka, 1973.
39. EMANUEL, N. M., ZAIKOV, G. E. and MAIZUS, Z. K., *Oxidation of Organic Compounds, Effect of Medium*, Oxford, Pergamon Press, 1984.
40. EMANUEL, N. M. and ZAIKOV, G. E., *Vysokomolek. Soed.*, **10A** (1968), 1475.
41. ZAIKOV, G. E., *Uspekhi Khimii*, **44** (1975), 1805.
42. BERLIN, AL. AL. and ENIKOLOPOV, N. S., *Vysokomolek. Soed.*, **A10** (1968), 1475.
43. IVANOVA, L. V., MOISEEV, YU. V. and ZAIKOV, G. E., *Izvestiya AN SSSR, ser. khim.* (1970), 2501.
44. IVANOVA, L. V., MOISEEV, YU. V. and ZAIKOV, G. E., *Vysokomolek. Soed.*, **A14** (1972), 1057.
45. IVANOVA, L. V., MOISEEV, YU. V. and ZAIKOV, G. E., *Vysokomolek. Soed.*, **A16** (1974), 1831.
46. IORDANSKII, A. L., MOISEEV, YU. V. and ZAIKOV, G. E., *Vysokomolek. Soed.*, **A14** (1972), 801.
47. IORDANSKII, A. L., MOISEEV, YU. V. and ZAIKOV, G. E., *Vysokomolek. Soed.*, **A16** (1974), 849.
48. MARKIN, V. S., ZAIKOV, G. E. and MOISEEV, YU. V., *Vysokomolek. Soed.*, **A12** (1970), 2174.
49. ZAIKOV, G. E., IORDANSKII, A. L. and MARKIN, V. S., *Diffuziya Elektrolitov v Polimerakh*, Moscow, Khimiya, 1983.
50. MANIN, V. N. and GROMOV, A. N., *Phiziko-khimicheskaya Stoikost' Polimernykh Materialov v Usloviakh Ekspluatatsii*, Leningrad, Khimiya, 1980.

51. ARTSIS, M. I., CHALYKH, A. E. and ZAIKOV, G. E., *Vysokomolek. Soed.*, A15 (1973), 63.
52. IORDANSKII, A. L., MOISEEV, YU. V., ZAIKOV, G. E., *Vysokomolek. Soed.*, A16 (1974), 849.
53. IORDANSKII, A. L., MOISEEV, YU. V. and ZAIKOV, G. E., *Doklady AN SSSR*, 219 (1974), 316.
54. ARTSIS, M. I., CHALYKH, A. E. and ZAIKOV, G. E., *Vysokomolek. Soed.*, A17 (1975), 128.
55. IORDANSKII, A. L., MOISEEV, YU. V. and ZAIKOV, G. E., *Vysokomolek. Soed.*, A16 (1974), 2248.
56. IORDANSKII, A. L. and ZAIKOV, G. E., *Diffuziya Nash Drug i Nedrug*, Moscow, Znaniye, 1982.
57. SHTERENZON, A. L., REITLINGER, S. A. and TOPINA, L, P., *Trudy III Mezhdunarodnogo Kongressa po Korrozii Metallov*, Moscow, Mir, 1968, pp. 130-41.
58. BERSHTEIN, V. A. and EGOROVA, L. M., *Vysokomolek. Soed.*, A19 (1977), 1260.
59. NECHAEV, P. P., MOISEEV, YU. V. and ZAIKOV, G. E., *Vysokomolek. Soed.*, A14 (1972), 1048.
60. RAZUMOVSKII, L. P., MOISEEV, YU. V. and ZAIKOV, G. E., *Vysokomolek. Soed.*, A19 (1977), 1358.
61. MOISEEV, YU. V. and ZAIKOV, G. E., *17th Symposium on Macromolecules*, Prague, 1977, p. 27.
62. RUDAKOVA, T. E., MOISEEV, YU. V. and ZAIKOV, G. E., *Vysokomolek. Soed.*, A16 (1974), 1356; A14 (1972), 449; A17 (1975), 1797.
63. PYATAKINA, N. K., MOISEEV, YU. V. and ZAIKOV, G. E., *Vysokomolek. Soed.*, A13 (1971), 200.
64. ARTSIS, M. I. and CHALYKH, A. E., *Tesisy Doklada ha Vsesoyuznoy Konferentsii po Vysokomolekulyarnym Soedineniyam*, Kazan, 1973, pp. 17-18.
65. IVANOVA, L. V., *Tesisy Doklada nd Konferentsii po Khimii Atsetaley*, Frunze, 1971, p. 5.
66. ALLOCK, H. R., *J. Macromol. Sci.*, C, 4 (1970), 149.
67. METHIN, I. A., PIOTROVSKII, K. B. and YUZHELEVSKII, YU. A., *Zhurnal Prikladnoy Khimii*, 48 (1975), 1108.
68. MINSKER, K. S., KOLESOV, S. V. and ZAIKOV, G. E., *Destruktsiya i Stabilizatsiya Polimerov na Osnove Polivinilkhlorida*, Moscow, Nauka, 1982.
69. RUDAKOVA, T. E., MOISEEV, YU. V., PORCHKHIDZE, A. D., KAZANTSEVA, V. V. and ASKADSKII, A. A., *Vysokomolek. Soed.*, A22 (1980), 449.
70. PORCHKHIDZE, A. D., RUDAKOVA, T. E., MOISEEV, YU. V., KAZANTSEVA, V. V. and ASKADSKII, A. A., *Vysokomolek. Soed.*, A22 (1980), 783.
71. RUDAKOVA, T. E., MOISEEV, YU. V., PORCHKHIDZE, A. D. and ZAIKOV, G. E., *Sonderdruck aus Kunstoffe-Fortschrittsberichte*, Vol. 5, Munich-Vienna, Carl Hanzer Verlag, 1979, pp. 25-32.
72. TIKHOMIROVA, N. I., KAZANTSEVA, A. A., TEYES AKUNYA, G., RUDAKOVA, T. E. and ASKADSKII, A. A., *Vysokomolek. Soed.*, A20 (1978), 1543.
73. ASKADSKII, A. A., *Deformatsiya Polymerov*, Moscow, Khimiya, 1973.
74. ZAIKOV, G. E. and RAZUMOVSKII, S. D., *Vysokomolek. Soed.*, A23 (1981), 513.
75. RUDAKOVA, T. E., ZAIKOV, G. E. and MOISEEV, YU. V., *Vysokomolek. Soed.*, A17 (1975), 1791.

76. PASHKYAVICHUS, V. V., MOISEEV, YU. V. and ZAIKOV, G. E., *Nauchnoissle-dovatel'skiye Trudy Lit. NII Tekstil'noy Promyshlennosti*, **3** (1974), 165.
77. MACHULIS, A. N. and TORNAU, E. E., *Diffusionnaya Stabilizatizatsiya Polymerov*, Vilnus, Mintis, 1974.
78. ZAIKOV, G. E., MARKIN, V. S. and IORDANSKII, A. L., *Plasticheskiye Massy* (1982), 38.
79. ZAIKOV, G. E. and NECHAEV, P. P., *Vysokomolek. Soed.*, **A26** (1984), 3.
80. RAZUMOVSKII, L. P., MARKIN, V. S. and ZAIKOV, G. E., *Vysokomolek. Soed.*, **A26** (1984), 5.

Chapter 4

CHARACTERISATION OF OXIDISED POLYOLEFINES BY REACTION WITH SULPHUR DIOXIDE

T. J. HENMAN

Cambridge Polymer Consultants, Melbourn, Herts, UK

SUMMARY

Oxidative processes are responsible for the vast majority of polyolefine degradation problems and the quantitative estimation of oxygenated groups is therefore of vital importance. Traditional methods based on carbonyl index or oxygen absorption have many disadvantages and sensitive techniques have been developed based on reactions with sulphur dioxide; these give insights regarding the sites of oxidative attack. The formation of sulphur-containing groups is considered in relation to attack on polyolefines by sulphuric acids and the mode of action of sulphur-containing antioxidants.

1. INTRODUCTION

The opportunities for oxidative reactions are best considered by reference to polymerisation processes (see Scheme 1).

The following abbreviations are used throughout:

PE = LDPE—low density polyethylene (high pressure process)
HDPE—high density polyethylene
PP—polypropylene

Oxidation will occur if unstabilised polymer is exposed to gases containing even parts per million levels of oxygen for prolonged periods above about 80 °C, and over the course of months or years at ambient temperatures.

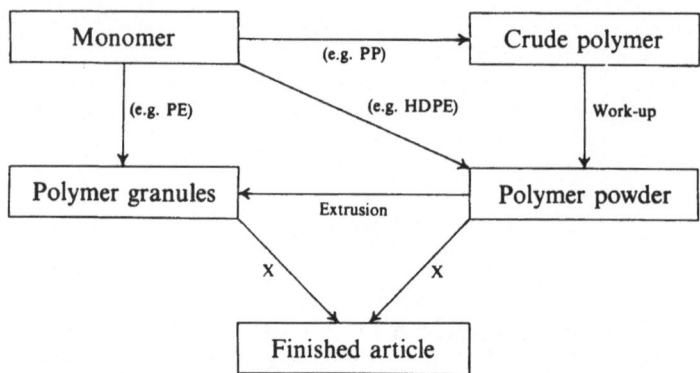

SCHEME 1. Polyolefine polymerisation processes. (X = a processing step, e.g. injection moulding.)

Although virtually all of the work described refers to LDPE/PP, there is no reason to suppose that most methods cannot similarly be applied to other polyolefines, e.g. HDPE, poly(4-methylpentene-1) and polybutene. In the case of ethylene–vinyl acetate (EVA) and related copolymers, the carbonyl group will of course interfere with conventional measurements.

Although there is a possible risk of oxidation during work-up (e.g. drying), granulation, silo storage and whilst awaiting conversion to the finished article, by far the most important opportunity occurs during service, where polymer is exposed to oxygen in the presence of heat, u.v. radiation, etc. The particular combination of factors during extrusion which affect the melt stability of PP has been discussed previously.[1]

The published literature in this field is vast, probably amounting to several thousand references, and drastic selection has been necessary to keep this chapter within manageable proportions. Polyolefine oxidation is of course greatly influenced by the presence of antioxidants, but this aspect has been excluded, since it is well covered in a recent book.[2] Similarly, much work has been done on atactic PP, a system favoured by many groups since homogeneous reactions can be studied. However, in the writer's opinion, work on atactic PP or so-called 'model hydrocarbons' has little relevance to the problems of the oxidation of isotactic PP and such work will only be mentioned briefly here. Finally, much work has been done, especially on PP, using initiated oxidation (e.g. by added peroxides, refs 3 and 4). Commercial application to make narrow molecular weight distribution PP (so-called 'controlled rheology' grades) is now well established, but the degradation products of such initiators must be taken into account during any characterisation.

2. CONSEQUENCES OF POLYOLEFINE OXIDATION

This reaction is catalysed by its products, a situation known as auto-catalysis leading to a parameter defined as the induction period (Fig. 1). Polyolefines suffer mechanical damage at relatively low levels of absorbed oxygen, e.g. $0.20/0.06\%$ (as carbonyl) for PE/PP, respectively. At this stage, crosslinking and/or degradation will have caused a considerable change in molecular weight distribution. If photoprocesses have been involved, there may also be chalking, cracking, delamination, increased dirt retention, etc. At one fifth of these levels there will be pronounced colour and odour changes caused by the formation of both carbonyl and unsaturation:

$$RH \longrightarrow RO_2H \longrightarrow \begin{cases} RCOR \\ RCHO \\ RO_2H \\ ROH \end{cases} \longrightarrow \begin{cases} \text{Esters} \\ \text{Lactones} \\ \text{Conjugated olefines} \\ \text{etc.} \end{cases}$$

However, the consequences of oxidation even at 30–100 ppm oxygen absorbed are very important. Differences can be seen in odour, seal

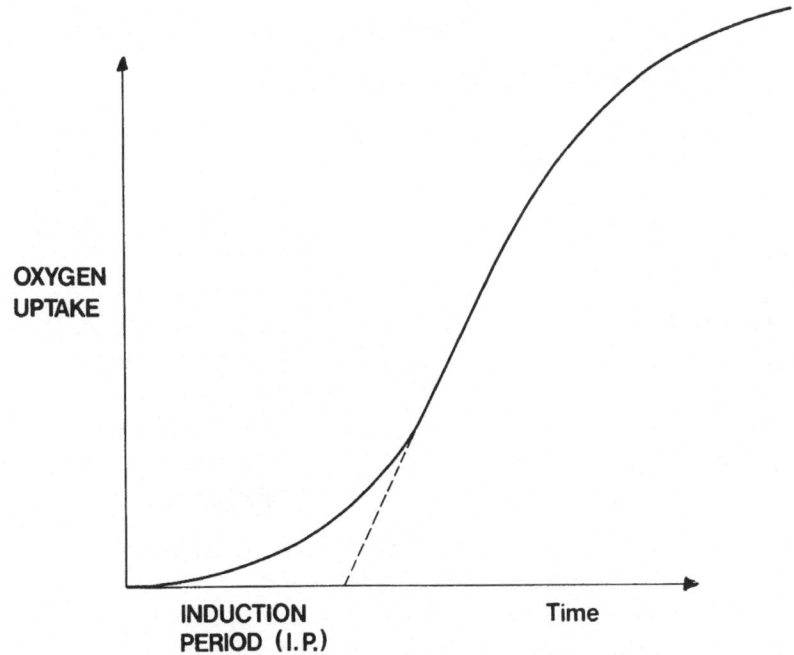

OXYGEN UPTAKE

INDUCTION PERIOD (I.P.)   **Time**

FIG. 1. Typical oxygen uptake curve.

strength, storage stability, etc., and it is in this region that the basic oxidation process has been studied.

3. MECHANISMS OF POLYOLEFINE OXIDATION

3.1. General

Several polyolefines, and PP in particular, would never have become commercial realities unless the mechanism of oxidation had been elucidated and methods developed for effective protection by combinations of antioxidants and u.v.-stabilisers. The standard free-radical oxidation mechanism[5] for hydrocarbon polymers, as shown in Scheme 2, is very familiar.

Initiation $\quad\quad RH \longrightarrow R^{\cdot} \quad\quad\quad$ (i)

Propagation
$$R^{\cdot} + O_2 \longrightarrow RO_2^{\cdot}$$
$$RO_2^{\cdot} + RH \longrightarrow RO_2H + R^{\cdot} \quad (ii)$$
$$RO_2H \longrightarrow RO^{\cdot} + OH^{\cdot} \quad (iii)$$
$$RO^{\cdot} \longrightarrow R^{\cdot} + {-}CO{-} \quad (iv)$$

Termination
$$2R^{\cdot} \longrightarrow RR$$
$$2RO_2^{\cdot} \longrightarrow RO_2R + O_2 \quad (v) \text{ For PP}$$
$$\searrow ROH + {-}CO{-} + O_2 \quad \text{For PE, etc.}$$

SCHEME 2.   Free radical polyolefine oxidation.

Clearly, for such a scheme to occur, there must be both oxygen source and initiator.

Oxygen source		*Initiating system*
Air		Photo—u.v.
Oxygen (triplet)		Photo—visible
Oxygen (singlet)		High energy (α, β, γ)
Ozone	plus	Corona discharge
Oxidising acids		Thermal
(e.g. H_2SO_4/HNO_3)		Mechanical (shear)
		Peroxides/hydroperoxides
		Biochemical

The key initiation step (i) is formation of an alkyl radical which can then react in a variety of ways as shown in Scheme 3.

Decomposition of the hydroperoxide to other species (iii) takes place under the influence of heat or light. PE[7] and PP[8-10] hydroperoxides

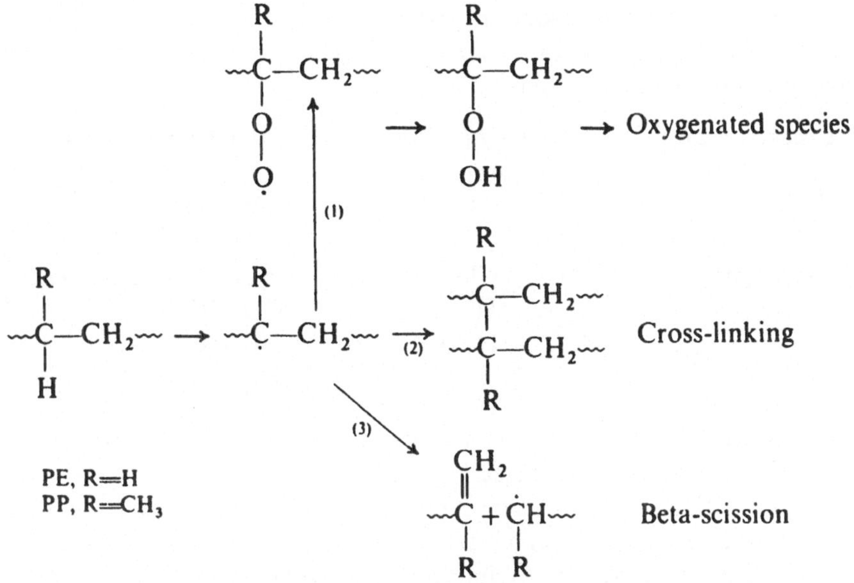

SCHEME 3.   Reactions of alkyl radicals ex polyolefines.

have been extensively studied, particularly by Chien and his co-workers. Functional group distribution is different for PE and PP with PE-OOH groups being isolated and PP-OOH groups mostly present in sequences of two or more.

3.2. Thermal Oxidation

A vast amount of work on the kinetics of polyolefine oxidation has been done by various Russian groups over the past 10 years. It is a mistake to believe that all the oxygen absorbed will be found as hydroperoxide since, apart from the termination reactions (v) yielding peroxides, intra-molecular chain transfer leads to radical isomerisations of hydroperoxide[11]

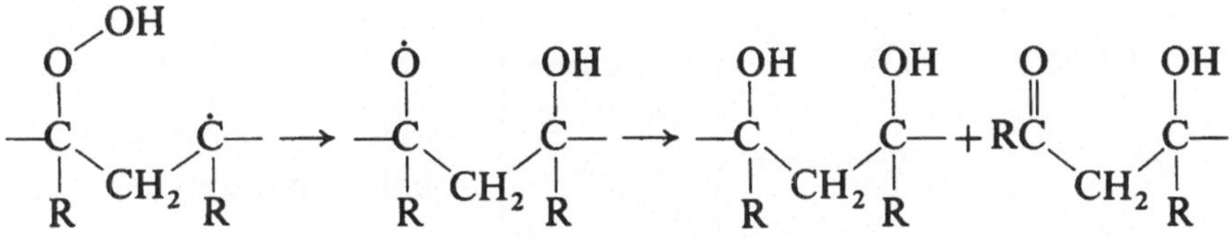

and the formation of hydroxyl and carbonyl groups.[12] Functional group content[13] of photo-oxidised polyolefines has been analysed[14] but results will be dependent upon the exact system used.

A range of volatile oxidation products is produced from the break-down[15] of hydroperoxides;[16] these have been analysed in detail[17] for oxidation at 150 °C. Products include:

'Cracked chains'	Elemental oxidation products	Chain oxidation products
Hydrocarbons	CO	CH_3OH
		$HCHO/CH_3CHO$
Hydrogen	CO_2	HCO_2H/CH_3CO_2H
	H_2O	CH_3COCH_3
	H_2O_2	Diones
		Esters, etc.

From Chien's work on rate constants, approximate half lives, $t_{1/2}$ of PP-OOH can be calculated (see Fig. 2).

Extrapolation to an extrusion temperature of 250 °C shows $t_{1/2} \sim 10$ s, that is, total decomposition within the residence time, whereas at 25 °C, $t_{1/2} > 2$ years is predicted. At intermediate temperatures, e.g. 150 °C, a common oven ageing temperature, a hydroperoxide half life of about 0·5 h confirms that non-radical hydroperoxide decomposers will be of benefit in this test but not necessarily at a service temperature of 50 °C (estimated $t_{1/2} = 40$ days).

In the oxidation of 50 µm PP film, Richters[18] compared infrared and MIR spectra and deduced that oxidation took place predominantly in the surface areas of the film. Hydroperoxide groups exist primarily[19] as blocks and in isolated positions and this ratio is affected by the oxygen pressure during their formation.[20] Thermal decomposition of PP hydroperoxide (both isotactic and atactic) was shown[10] to consist of two reactions. An initial fast reaction, probably an intramolecular radical-induced mechanism accounting for over 70 % of the total, is followed by a reaction more than 50 times slower, involving decomposition of β-hydroxyhydroperoxides formed as by-products of the first reaction. A variety of products is formed:[21]

A radical yield of 1·8 for hydroperoxide decomposition can be compared with a theoretical value of 2·0. Decomposition of PE hydroperoxide was

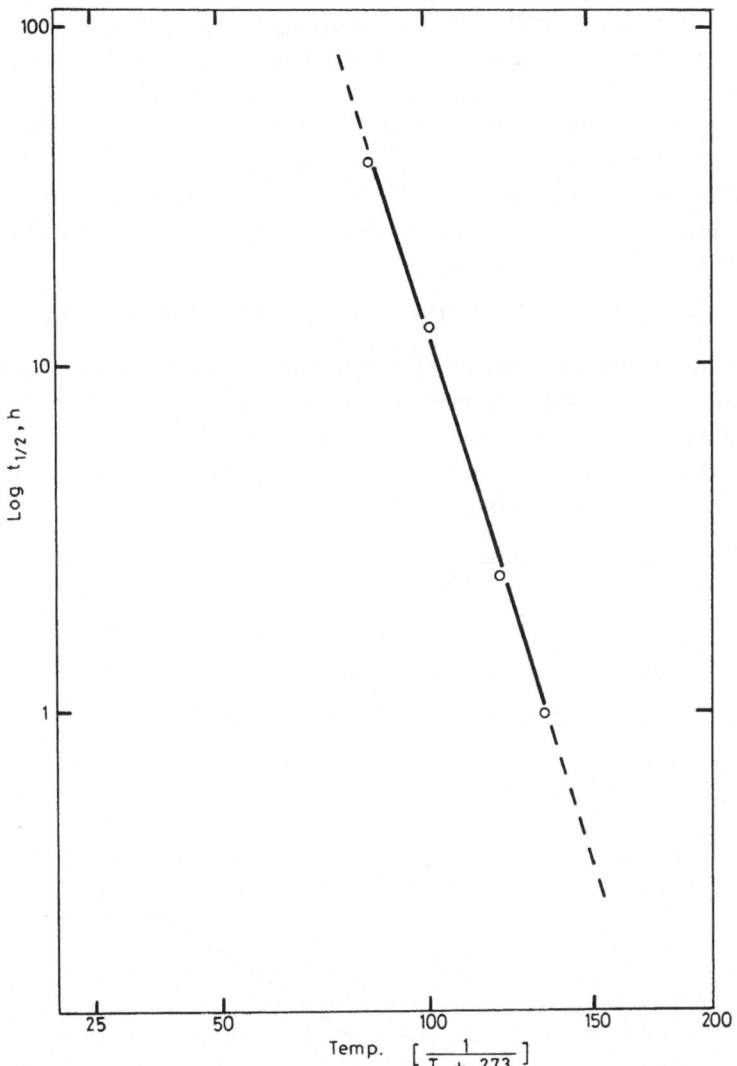

FIG. 2.   Half lives for PP-OOH decomposition.

studied by a Hungarian group,[22] who found that contrary to popular belief it is not a simple unimolecular process but is affected by carbonyl content.

3.3. Photo-oxidation
An already complicated picture becomes even more obscure. Whereas carbonyl concentration rises consistently, the position and maximum concentration of hydroperoxide depends very much on the irradiating

source, sample geometry, etc. Hydroperoxide level cannot be more than proof that oxidation has taken place. The infrared spectra of photo-oxidised samples are generally considerably more complicated than those from thermal oxidation but some clues are characteristic, for example, the disappearance of vinylidene[23] in PE and the appearance of vinyl[24] in PP. Detailed photo-oxidation chemistry is best discussed in terms of Norrish type I and II processes.[2] If the latter occurred exclusively, equal amounts of carbonyl and vinyl would occur:

$$R-CO-CH_2CH_2CH_2R' \longrightarrow RCOCH_3 + CH_2=CHR'$$

In Fig. 3 the development of functional groups in PE subjected to a screened medium pressure mercury arc is shown. Up to $c. 0.05\%$ O_2 (as carbonyl), absorbed, carbonyl and vinyl groups are formed in equal proportions but above that level it would appear that type I processes (producing carbonyl + alkyl radicals) become more important.

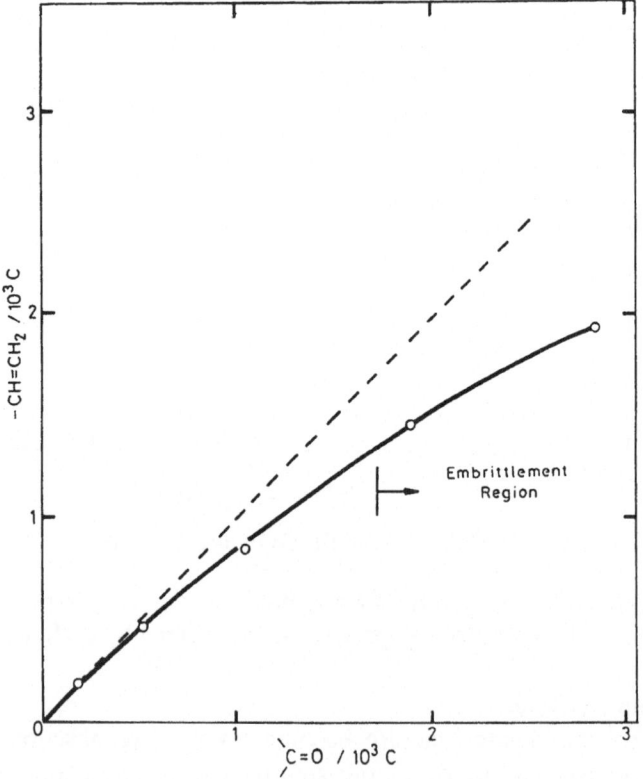

FIG. 3. Correlation of vinyl/carbonyl development in photo-oxidised PE.

3.4. Radiation-induced Oxidation

The radical processes induced by high energy radiation are essentially random in character. Although some radical redistribution may take place during the post-irradiation period, when the primary radicals react with molecular oxygen, hydroperoxides tend to be isolated rather than in groups. The consequent lower extent of hydrogen bonding in PP therefore leads to a decomposition rate $> 50\%$ higher.[12]

3.5. Chemical Oxidation

Oxidation of polyolefines with 'chemical' reagents such as sulphuric[25] or nitric[26] acids must be included for completeness. However, reactions are essentially ionic in character and, as in all polymers, polymer crystallinity plays an important role. Reactions are mainly confined to surface layers, and can therefore be studied by ESCA. Cameron and Main[25] showed that PE was sulphonated, initially at the branch points; the product had new absorptions at 1160, 1040, 580 cm^{-1}, consistent with the formation of —SO$_3$H groups. After heating at 100 °C, the 1160 cm^{-1} band was partially converted into an olefine and the following mechanism was proposed:

$$
\begin{array}{c}
R \\
| \\
-CH_2-CH-CH_2- \\
\\
\downarrow\, H_2SO_4 \\
\\
R \\
| \\
-CH_2-C-CH_2- \\
| \\
OSO_2OH \\
\\
\downarrow\, \text{Heat, 100 °C} \\
\\
R \\
| \\
-CH_2-C{=}CH- + H_2SO_4
\end{array}
$$

3.6. Computer Modelling

Computer modelling of the thermal oxidation of PP appeared in the early 1970s.[27] Within the past year or so, two groups, one in Canada[28] and one in Hungary,[29] have reported computer modelling of PE photo-oxidation. Over 50 individual reactions have been defined and lifetimes of unstabilised

PE predicted[30] to be in the range 3 months–2 years, depending upon service temperature.

3.7. Effect of Other Components

Space does not allow a discussion of the important effect that metal ions can have on oxidation processes. The reader is best referred to recent reviews covering photoinitiators,[2] metals (especially copper)[31] and pigments.[32]

4. METHODS FOR ASSESSMENT OF OXIDATION

4.1. Direct Measurement

It is unfortunate that there is no suitable radioisotope of oxygen[33] (^{15}O has a $t_{1/2} < 0.01$ years) since this would provide a splendid vehicle for oxidation studies. Apart from activation analysis,[34] the only absolute method is oxygen uptake by techniques such as the Barcroft apparatus (Fig. 4). Although very sensitive pressure transducers have been used,[35] the limitation is temperature control of the bath. Answers are necessarily 'integrated' results, and the technique is now used less often, although an automated apparatus has been described.[36]

4.2. Physical Measurements

Any such measurements (Table 1) illustrate the effect of oxygen previously absorbed; if the test is carried out at or above 150 °C ($t_{1/2}$ for

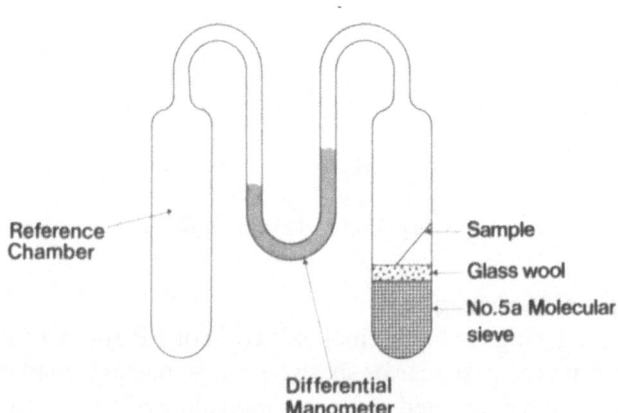

FIG. 4. Barcroft oxygen uptake apparatus.

TABLE 1
PHYSICAL METHODS FOR CHARACTERISATION OF OXIDATION

Technique	Suitable for samples		Sample form		
	Unstabilised	Stabilised	Powder	Granules	Film
O_2 uptake	×	×	×	×	×
DTA		×		×	×
MFI	×[a]	×	×	×	
MFI (prolonged)	×[b]	×[b]			×[b]
MFI $(SO_2)^g$	×[c]	×[c]	×	×	
MWD	×	×	×	×	×
Swell ratio[d]	×	×	×	×	
Colour	×	×		×	×[e]
Colour $(SO_2)^g$	×	×		×	×[e]
Odour[f]	×	×	×	×	×

[a] PP samples must be stabilised before measurement. [b] Only examined for PE granules. [c] Most suitable for PP. [d] For PE only—measure of crosslinking. [e] Only for roll of film unless microscopy used. [f] Highly subjective unless odour panel used. [g] Details discussed later in text.

$RO_2H = 0.5$ h), results will include the effects of any hydroperoxide decomposition taking place within the timescale of the test (see Section 3.2).

Conventional Melt Flow Index (MFI) tests[37] are carried out at 190 °C (PE/PP), 230 °C (PP) and sometimes 270 °C (PP) and are subject to about a 5–10 % error.

It has been found that a 'prolonged MFI' test is a good indicator of potential crosslinking problems in PE. A standard 190 °C/2·16 kg test uses a 5 min warm-up before the weight is released; a 'prolonged test' uses an additional 30 min warm up. A value for the ratio

$$\frac{\text{Standard MFI}}{\text{Prolonged MFI}} > 1 \cdot 1$$

indicates a possible oxidation problem.

Since sulphur dioxide destroys hydroperoxides, similar tests can be used for PE and PP to show the presence of oxidation (see Table 2). In both cases SO_2 treatment is at room temperature for >4 h or, more conveniently, overnight. As oxidation leads to reduction in MFI due to crosslinking for PE and an increase due to degradation in PP, the ratios are inverted; once again a value greater than 1·1 is significant.

TABLE 2
THE MFI (SO$_2$) TEST

Polymer	Test[a] load (kg)	Ratio used
PE	5	Treated/untreated
PP	10	Untreated/treated

[a] Standard test at 190 °C/5 min warm up but without the additional stabilisers normally used for PP.

4.3. Spectroscopic Methods

These are summarised in Table 3. In 1955 work at ICI showed that the weak infrared (i.r.) absorptions of the primary reaction product, hydroperoxide, occur[38] at c. 3520 cm^{-1} (free) and 3380 (hydrogen-bonded). The claim by Mitchell and Perkins[39] that 0·1 ppm OOH in PE should be detectable at 1195 cm^{-1} after reaction with SO$_2$ is highly suspect; this reaction is discussed in Section 5.3.

TABLE 3
SPECTROSCOPIC METHODS FOR ASSESSMENT OF OXIDATION

Group	Spectroscopic method	Other methods	Suitable for samples		Sample form		
			Unstabilised	Stabilised	Powder	Granules	Film
OOH	i.r.		×	×[a]			×
	i.r./SO$_2$		×	×			×
		Iodometry	×·	×	×	×[d]	×[c]
		SO$_2$—XRF	×	×[b]	×	×	×
OOR		Iodometry	×	×	×	×[d]	×[c]
CO	i.r.		×[g]	×[e]			×
	Derivative u.v.		×	×[a,e]			×
		DNP	×	×			×
C:C	i.r.		×	×[a]			×
	Derivative u.v.		×	×[a]			×
	i.r./Br$_2$		×	×			×
	i.r./SO$_2$		×	×			×
OH	i.r.		×	×[a]			×
Miscellaneous	Fluorescence[f]		×	×	×	×	×

[a] Phenolic antioxidants can interfere. [b] Thioesters, etc., can interfere. [c] Thin films only. [d] If total solution method used. [e] Some —CO— containing antioxidants will interfere. [f] Excited by long wave u.v., e.g. 300 nm. [g] Certain copolymers (e.g. EVA) will interfere.

TABLE 4
CARBONYL GROUPS IN PHOTO-OXIDISED PE
(after ref. 23)

Group	$v \ (cm^{-1})$
RCO$_2$H	1 710
RCOR′	1 720
RCOCH$_3$	1 725
RCHO	1 735
RCO$_2$R	1 750
RCO$_3$H/RCO$_3$R	1 785

The x-ray fluorescence test gives results accurate to ± 5 ppm sulphur—a blank of 5 ppm for PP samples can be explained by traces of sulphur in the propylene monomer (see Section 5 for detailed discussion).

The traditional method of assessing oxidation is the carbonyl index which monitors the build-up of non-volatile secondary oxidation products, e.g. in PP at $1710\,cm^{-1}$; clearly a wide range of commercial stabilisers[40] will interfere. More recently, second order derivative u.v. spectroscopy has been used.[41] In photo-oxidised PE a wide range of carbonyl functions has been recorded[23] (see Table 4).

Billingham[42] used the reaction of ketone and aldehyde groups with 2,4-dinitrophenylhydrazine (DNPH) to indicate the position of locally oxidised regions in PP.

Although the reaction is not quantitative, its advantage is a great increase in sensitivity from i.r. (carbonyl) to u.v./visible (conjugated system); an error of 10 % has been quoted.[43] Theoretically, since DNPH does not react with ester groups, such a reaction could be used to monitor EVA oxidation.[35]

Carbonyl analysis of ethylene copolymers (e.g. with vinyl acetate or acrylates)

$$\left[\begin{array}{c} -CH_2-CH- \\ | \\ R \end{array}\right]_n \qquad (R = O.COCH_3/CO.OCH_3)$$

will clearly cause problems. However, a 17 % vinyl acetate copolymer was found[44] to produce vinyl groups during photo-oxidation (but *not* during thermal oxidation) at almost exactly the same rate as a homopolymer (see Fig. 3) so this method can be used. The SO_2 method has not been investigated for ethylene copolymers.

$C≡C$ unsaturation is probably responsible for the colour of polyolefines at higher oxidation levels; at lower levels several different types can be identified (Table 5).

A series of bands can sometimes be identified in the ultraviolet, separated by $c. 30$ nm, representing different lengths of conjugation; however, this is more relevant to degraded PVC. Changes in the i.r. spectra after bromine treatment or catalytic hydrogenation give some measure of the unsaturation originally present. Loss of pendant methylene during photo-oxidation has been discussed:[23] its use as a highly sensitive indicator of PE oxidation is described in Section 5.6. Finally, individual oxidised polymer powder particles or granules show fluorescence which starts to appear reproducibly at the same level as a 10 % change in MFI ratio as discussed in Section 4.2.

4.4. Hydroperoxide Analysis
Determination of hydroperoxide in atactic PP presents no problems as homogeneous solutions can be obtained. Various colorimetric methods

TABLE 5
UNSATURATION IN PHOTO-OXIDISED PE (after ref. 8)

Group	Structure	$v \ (cm^{-1})$
Pendant methylene	$>C=CH_2$	885
End chain vinyl	$-CH=CH_2$	910
trans-Vinylene	$-CH=CH-$	1 645

have been published;[45] ideally, phenolic antioxidants should not interfere[46] and the method should be suitable[47] for nominally 'non-oxidised' atactic PP.

For PE or isotactic PP, an iodometric method is normally used; this has a lower limit[48] of c. 30 ppm OOH and can be used for oxidised films below 200 μm thickness.[49] Some sophisticated methods claim[11] to distinguish between hydroperoxides and dialkyl peroxides via temperatures of titration but the latter are not involved in the early stages of oxidation.

4.5. Miscellaneous Methods

A useful summary[50] has recently been updated[51] and detailed discussion is not justified here. In the majority of cases assessment of stabilised samples is required. Conventional methods for assessing oxidative stability (e.g. oven ageing, DSC, DTA, TGA) can usually be discounted due to lack of suitable standards. Similarly, by the time changes in mechanical properties[52,53] have occurred the sample is well on its way to failure. Given a radical mechanism, electron spin resonance (ESR)[54-56] and electron spin echo[57] have been used but clearly ignore non-radical products.

Since oxidation is principally a surface phenomenon, it is not surprising that surface techniques such as MIR/ATR[58,59] and ESCA[60] have been reported. A recent method[61] allows determination of the depth of oxidative penetration by cross-sectional polishing. Positron annihilation has been described;[62] however, few laboratories have positron generators.

The most encouraging development appears to be chemiluminescence, a technique apparently directly related to hydroperoxide functionality[74] and developed from earlier thermoluminescence[63] and luminescence spectroscopy[64] techniques. However, its great sensitivity means that care must be taken in the interpretation of results, but different aspects of this technique are currently being investigated in Australia,[65] USA,[66-71] Czechoslovakia[72] and the UK.[73]

5. REACTIONS OF OXIDISED POLYOLEFINES WITH SULPHUR DIOXIDE

5.1. General Observations

Most of the assessment methods described above are useful to different extents from about 100 ppm upwards, but are of little use in following the very small changes that occur during the induction period. In this initial stage of thermal oxidation it is believed[75] that hydroperoxide is formed almost quantitatively from the oxygen absorbed for both PE and PP, i.e.

$$RH + O_2 \longrightarrow RO_2H$$

5.2. Reaction Mechanism

Some clues as to the reaction mechanism are available in the literature. In 1953, dry SO_2 gas was found[76] to react explosively with cumene hydroperoxide to give phenol and acetone via an ionic mechanism even at $-55\,^{\circ}C$. The reaction was further investigated[77] 20 years later as a low temperature polymerisation system for vinyl chloride; in the presence of water, the mechanism switches to a free radical type. The maximum monomer conversion is obtained at a $[SO_2]/[OOH]$ ratio of $1:5$. In the presence of alcohols or ketones, SO_2 groups are incorporated into the polymer chain, and intervention of SO_3 radicals is shown by end group formation. The nature of radical intermediates was identified by ESR.[78,79]

Alkoxy radicals can cleave to give a ketone and an alkyl radical which can react further to give C–S bonds; the formation of RSO_2^- rather than $ROSO_2^-$ is favoured in the presence of water.

Such a scheme would provide a rationale for the production of a wide variety of sulphur-containing species as well as a series of compounds with S–S bonds which one would expect to be less thermally stable.

Other workers[80] believe that the homolysis of hydroperoxides occurs via perester formation at low SO_2 concentrations (10^{-4} M):

TABLE 6

PHYSICAL PARAMETERS OF POLYMER SAMPLES

Sample type	Parameter	Minimum path length (μm)
Powder particle	Diameter $\simeq 200\,\mu m$	100
Polymer granule	Size, e.g. $3 \times 5\,mm$	1 500
Pressed film	Thickness $200\,\mu m$	100

Above 10^{-3} M proton acids are said to cause heterolytic decomposition.

Recent model studies by the group at Aston[81] suggest that when the SO_2 is present in molar excess ($[SO_2]/[ROOH] > 1$), the true decomposition catalyst is SO_3, formed via a persulphonyl radical

$$ROO^. + SO_2 \longrightarrow ROOSO_2^. \longrightarrow RO^. + SO_3$$

whereas at low SO_2 ratios, radical processes predominate.

In polymeric systems the physical parameters have to be considered (Table 6).

5.3. Gas Diffusion

Diffusion of gases,[82] e.g. SO_2,[83] has been intensively studied in PE. Although oxidation can theoretically take place throughout any samples, its rate will clearly be diffusion controlled by the ingress of oxygen (see Table 7); thermal oxidation of solid PP is said[84] to be limited to a depth of $0.635\,nm$. Subsequent reaction with SO_2 will give conditions of SO_2 excess at the surface (SO_3-induced ionic process) and SO_2 shortage below the surface (radical process) until equilibrium has been established.

TABLE 7

DIFFUSION COEFFICIENTS FOR O_2/SO_2 IN PE/PP

Polymer	Diffusion coefficients[b] $\times 10^{10}$ ($cm^2\,s^{-1}$)	
	O_2	SO_2
PE	5 300	30^a
PP	4 700–6 600	3.5

[a] Decreases slightly as relative humidity increases—maximum i.r. absorbance was obtained[11] after c. 15 min for a $200\,\mu m$ thick PE sample.
[b] From refs.[85,86]

TABLE 8

EFFECT OF WATER ON THE SO_2 REACTION

Polymer	100°C oxidation time (h)	Sulphur (ppm) by x-ray fluorescence[a]	
		Dry[b]	Wet
PE	26·5	250	320
PP	7·5	400	880

[a] After vacuum treatment 50°C/6 h.
[b] 'Dry' = dried over P_4O_{10}.
  'Wet' = in presence of water droplets.

5.4. Effect of Water

The Aston work[81] involving SO_3 shows that it becomes a less effective hydroperoxide-decomposer as it reacts with water, either formed during the reaction or by dehydration of alcohols. One would therefore expect to see differences in the SO_2 reaction in the virtual absence and in the presence of water. Table 8 shows an increased retention of sulphur during 'wet' conditions, so control of reaction conditions is necessary for routine work.

5.5. Polypropylene

5.5.1. Thermal Oxidation

Our interest in reactions with SO_2 started in 1969 when we investigated the MFI change of a PP batch that had been exposed to very high storage temperatures for a prolonged period. We knew of the then recent paper,[39] which stated that 'with slight modification', the PE–SO_2 method 'should be suitable for determination of hydroperoxides in PP and other polyolefines'. The mechanism of the PP–SO_2 reaction has been discussed by Richters[18] who reported the brown colour formed in 50 μm PP film by decomposition of sulphates after SO_2 treatment. Localised brown spots were ascribed to the formation of conjugated double bonds by loss of sulphuric acid

$$\begin{array}{ccc} & CH_3 & CH_3 \\ & | & | \\ -CH\!\!=\!\!C\!\!-\!\!CH\!\!=\!\!C- \end{array}$$

and were found to have iron particles within them; spots could be seen 'long before the end of the induction period'.

In 1969 an exhaustive study[87] of the oxidation of diluent phase PP* was

* Diluent phase PP refers to the conventional process used since the late 1950s in which propylene is polymerised in the presence of a hydrocarbon diluent. It is increasingly being replaced by more cost effective 'gas phase' PP made in the absence of such diluents.

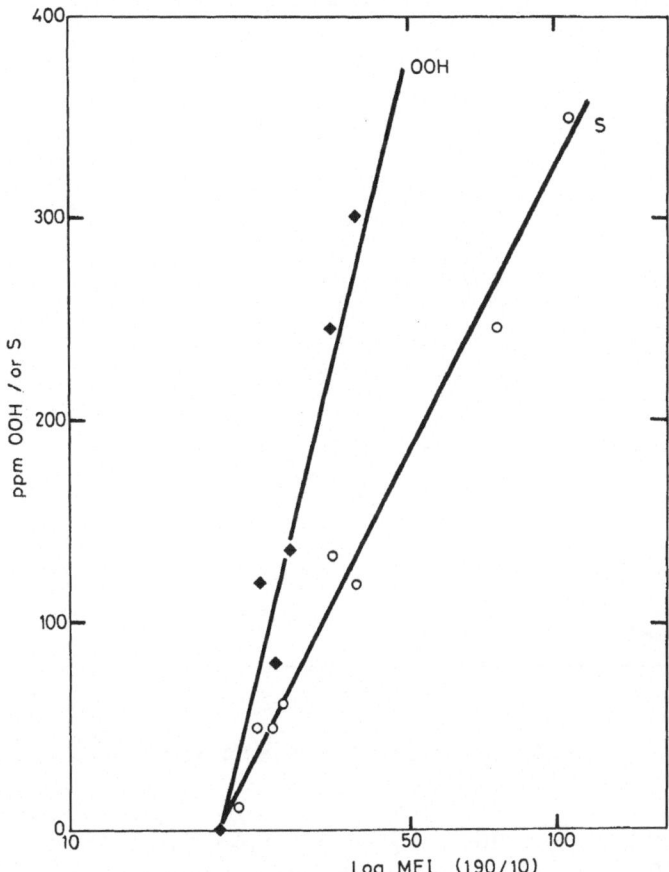

FIG. 5. PP oxidation—correlation of OOH/S with MFI.

undertaken. As shown in Fig. 5, it was found that for unstabilised PP powder oxidised at 100 °C for up to 36 h, log MFI plotted against hydroperoxide (via iodometry) or sulphur content (by XRF after SO_2 treatment) gave reasonable correlations. Competing processes occur because after an induction period of $c.$ 15 h hydroperoxides are formed almost exponentially but have a half life of $c.$ 12 h at this temperature.

Direct correlation (Fig. 6) of OOH versus retained sulphur (by XRF) showed a linear correlation up to 300 ppm OOH, with appreciable sulphur loss. Three results obtained in 1982[88] by ^{35}S analysis are included for comparison.

Examination of sulphur losses at room temperature on PP powder was extended to compression moulding temperatures (200–270 °C—Fig. 7).

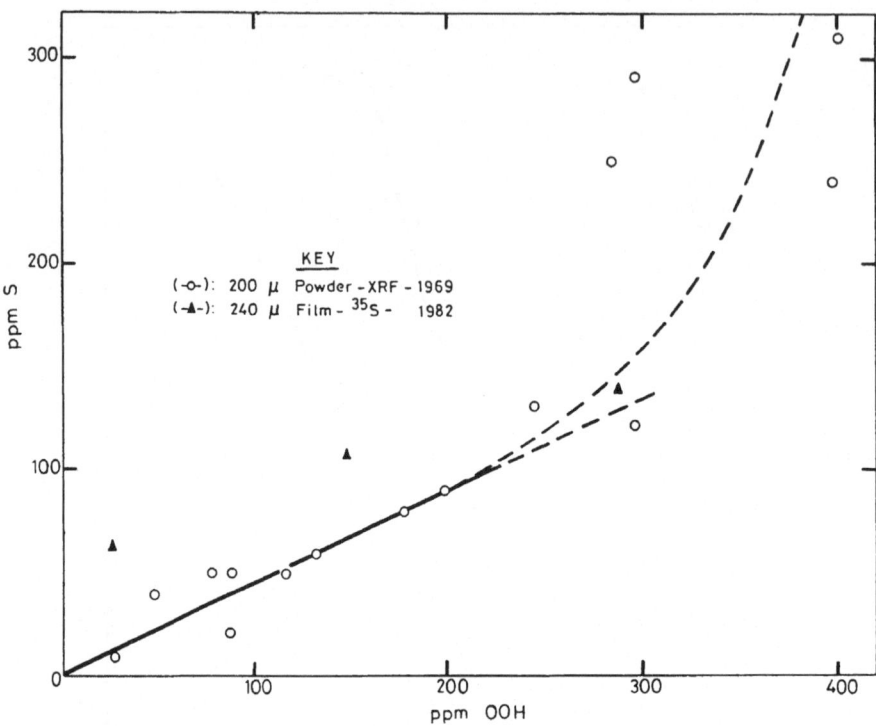

FIG. 6.   Correlation of OOH/S for oxidised PP.

For quantitative measurement granule samples need pressing into a disc before XRF measurement, when substantial additional losses are seen.

To summarise, therefore, the SO_2 technique is principally useful for very low OOH levels to show up differences in samples not resolvable by direct hydroperoxide analysis.

5.5.2. Photo-oxidation

The effect of SO_2 treatment on the rate of photo-oxidation[41] of unoxidised and oxidised diluent phase PP film is shown in Fig. 8. It is clear that prior treatment with SO_2 stabilises the polymer, doubling the induction period from 100 to 200 h. It is also interesting to note that both the pre-oxidised films (130 °C/0·5 h and 100 °C/4 h) are also markedly stabilised. In fact, the photostabilities of control and oxidised films are virtually the same after SO_2 treatment. It is evident from these results that although all the hydroperoxides—even those generated thermally—have been destroyed, photo-oxidation of the polymer still occurs, although at a later stage.

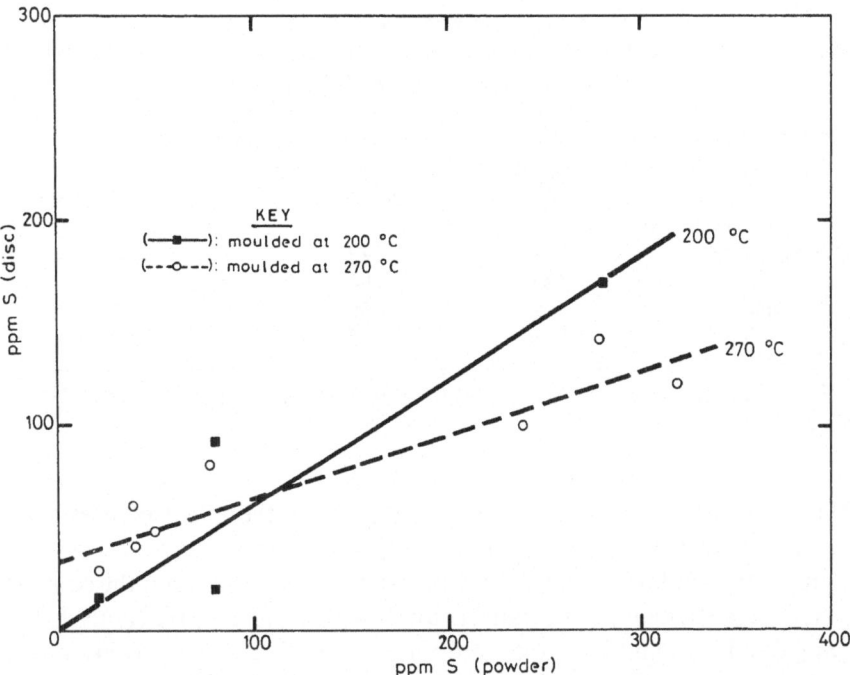

FIG. 7. Oxidised PP—loss of S on compression moulding of powder into discs.

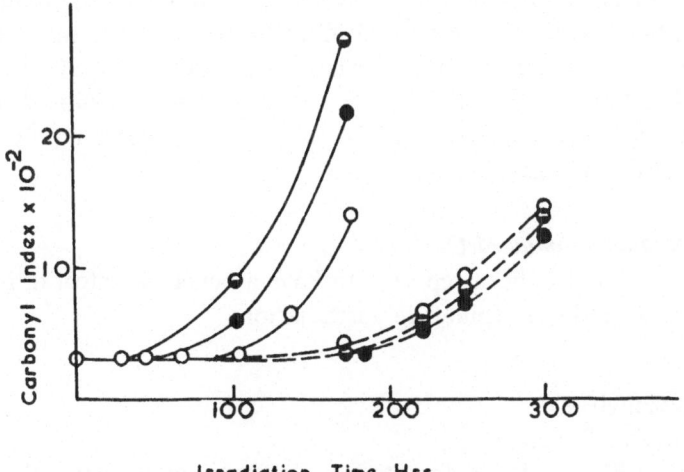

Irradiation Time, Hrs

FIG. 8. Rate of photo-oxidation of polypropylene film in a Microsal unit ($\sim 200\,\mu$m thick). ($\bigcirc$) Control; ($\ominus$) thermally oxidised for $\frac{1}{2}$h at $130\,^\circ$C; ($\bullet$) thermally oxidised for 4h at $100\,^\circ$C before (——) and after (---) SO_2 treatment.

TABLE 9

EFFECT OF SO_2 TREATMENT ON THE PHOTO-OXIDATIVE STABILITIES OF POLYPROPYLENE
FILMS BEFORE AND AFTER OVEN-AGEING

Polymer type	Oven ageing[b]	u.v. embrittlement times,[c] t (h)	
		Untreated	With SO_2 treatment
Diluent phase	No	140	220
	Yes	90	240
Gas phase[a]	No	100	70
	Yes	65	65

[a] Experimental sample. [b] 130°C/0·5 h. [c] 500 W Microscal unit. t = time to 0·06 %
O_2 as carbonyl.

This observation confirms earlier findings[89] on the effect of destroyed
hydroperoxides by photolysis in an inert atmosphere.

In PP the oxidation chemistry is markedly dependent upon the catalyst
system, a particularly important factor now that more active catalysts are
being developed and increasing proportions of PP are made by gas phase
processes.

The effects of SO_2 treatment[123] on the photo-oxidative stabilities of PP
films made by different polymerisation processes are compared in Table 9.
For diluent phase polymer, prior SO_2 treatment substantially improves
photostability, control and oxidised films having virtually the same light
stability after SO_2 treatment. In the case of gas phase PP, SO_2 treatment
had little effect on light stability; in fact, whereas the oxidised polymer
remained virtually unaffected by SO_2 treatment, the control polymer was
somewhat destabilised.

5.5.3. Radiation-induced Oxidation
Recently[90] SO_2 has been reported to have a beneficial effect in reducing
oxidative degradation following γ-irradiation.

5.6. Polyethylene
5.6.1. 'Defect Structures'
Whereas in PP the only 'defect structure' of any significance is the olefinic
end group, in PE there is a much wider variation,[91] since residues arising
from degradation of peroxide catalyst must also be considered[92] (see
Table 10).

TABLE 10
'DEFECT STRUCTURES' IN PE ($M_w \simeq 7 \times 10^5$)

Group	Number per 10^3 C
Branches (C_2–C_5)	20
Branches ($\geq C_6$)	3
Vinylidene	0·27
Vinylene/vinyl	0·08
Carbonyl	0·02

Although by analogy with PP one might expect the branch points to be very susceptible to oxidative attack, the reactivity of allylic hydrogens is very well established[93] and it is now generally accepted that initial oxidative reaction takes place adjacent to the vinylidene[94]

$$-CH_2-\underset{\underset{CH_2}{\|}}{C}-$$

or 'pendant methylene' units; this is of particular importance when reactions with SO_2 are considered. PE samples known to have low levels of oxidation show negligible differences in the infrared (Figs 9 and 10) until treated with SO_2 (Figs 11 and 12) when a reduction in pendant methylene absorption is seen; this can be further intensified by prior heating at 100 °C. These spectra were recorded using the FTIR technique.

5.6.2. Thermal Oxidation
Although earlier work[95,96] had shown the formation of bisulphates (**I**) Mitchell and Perkins[39] believed that the sulphate (**II**) is formed by comparison with dilauryl sulphate (**IIa**); this has been confirmed by more recent work.[97] X-ray photoelectron spectroscopy has given conflicting results, one paper[98] giving 'positive identification' of the bisulphate (**I**), 'the first direct evidence for this mechanism' and other work[99] finding evidence

$$ROOH \xrightarrow{SO_2} ROSO_2OH \xrightarrow{ROH} ROSO_2OR$$

$$\quad\quad\quad\quad\quad\; \textbf{I} \quad\quad\quad\quad\quad \textbf{II}$$

$$\textbf{Ia} \quad R{=}C_{12}H_{25} \quad \textbf{IIa}$$

$$\begin{array}{c} CH_2O \\ CH_2 \diagup \quad \diagdown SO_2 \\ CH_2O \diagup \\ \textbf{III} \end{array}$$

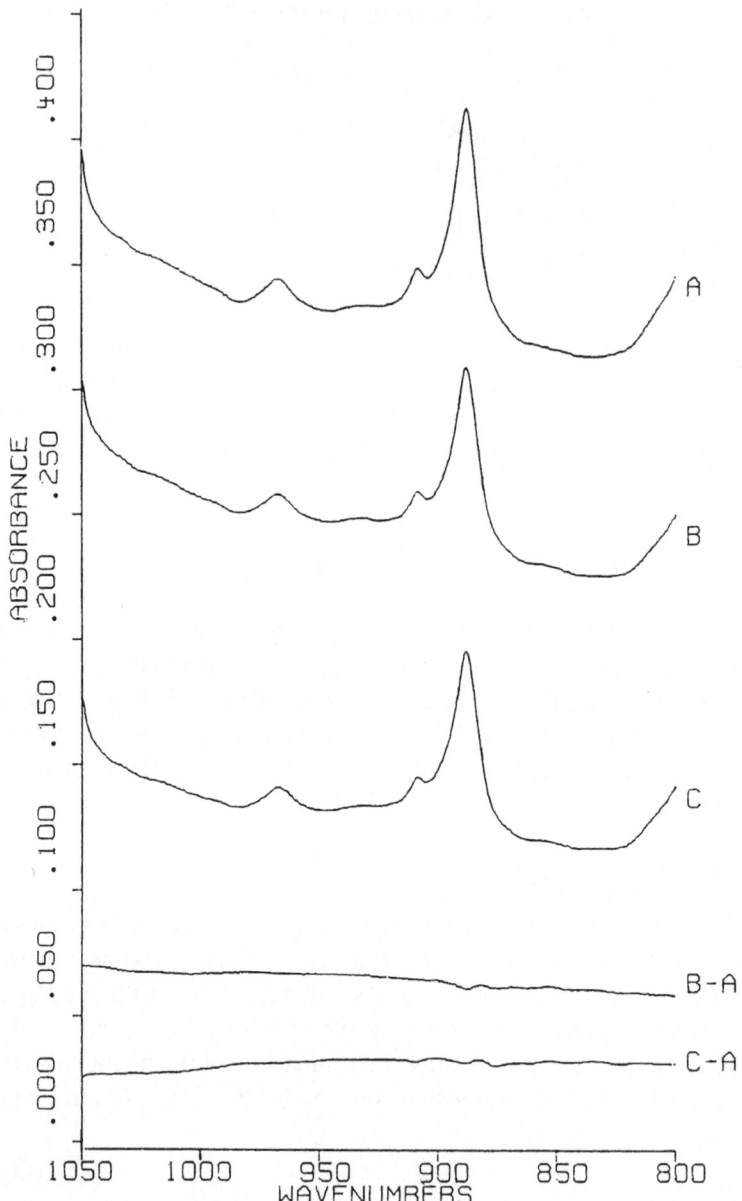

FIG. 9. Samples as received, LDPE spectra ($800-1050\,cm^{-1}$). A, Control;
B, oxidised sample 1; C, oxidised sample 2.

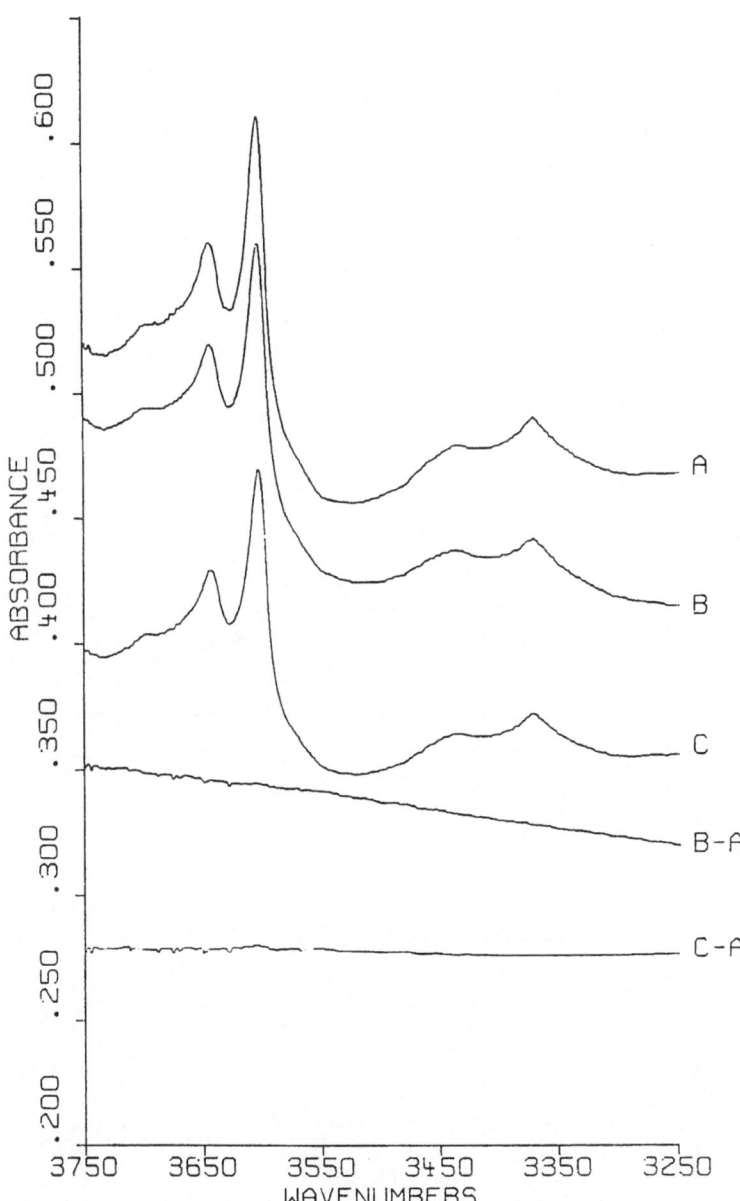

FIG. 10.   Samples as received, **LDPE** spectra (3250–3750 cm^{-1}). A. Control;
B, oxidised sample 1; C, oxidised sample 2.

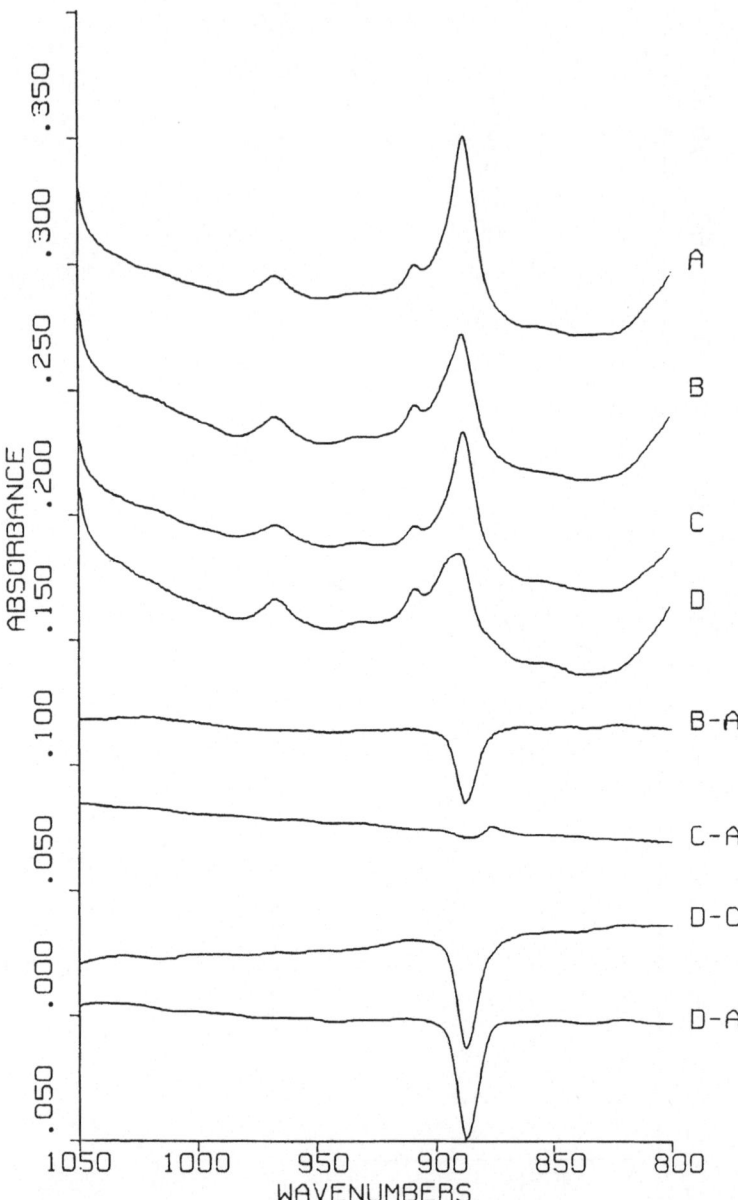

FIG. 11. LDPE spectra: effect of SO$_2$. Oxidised sample 2. A, As received; B, after SO$_2$ treatment (60 ppm S retained); C, heated 100°C/18 h; D, heated 100°C/18 h, then SO$_2$ treated.

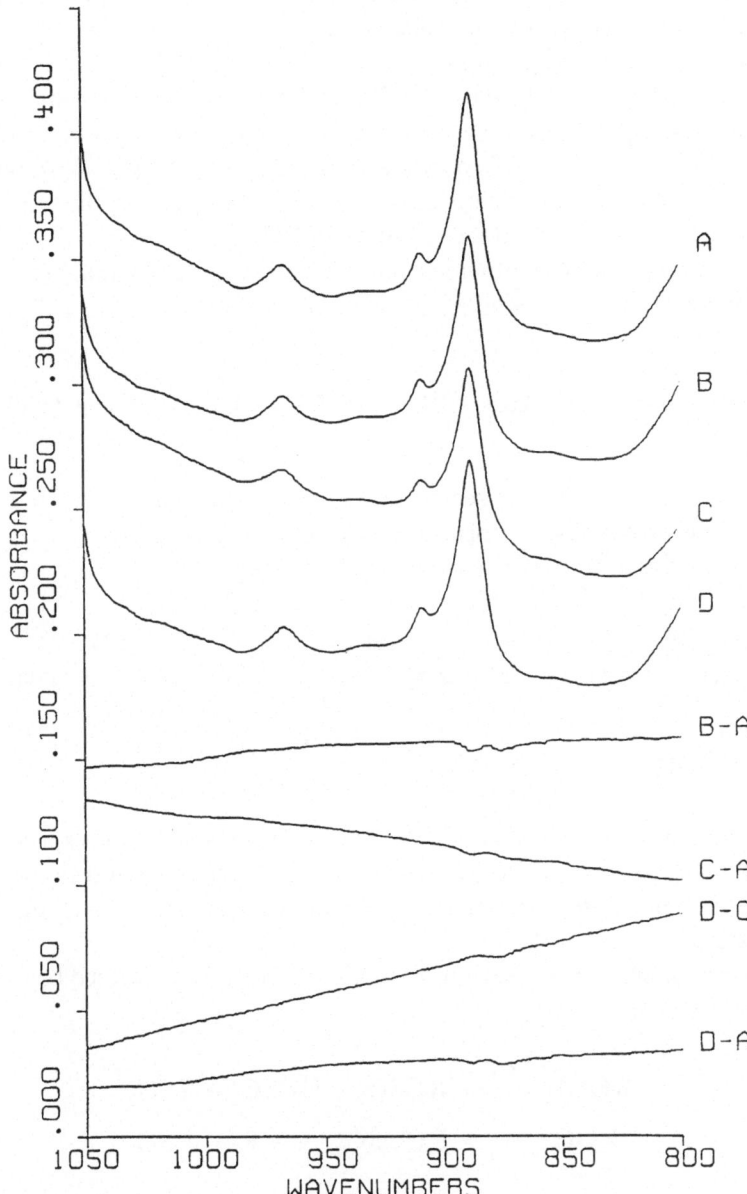

FIG. 12. LDPE spectra: effect of SO_2. Control. A, As received; B, after SO_2 treatment (30 ppm S retained); C, heated 100 °C/18 h; D, heated 100 °C/18 h, then SO_2 treated.

for both $C-O-SO_2$ (sulphate/bisulphate) and $C-SO_2$ (sulphone) groups at the surface. However, as noted earlier (Section 3.2) the types and environments of OOH groups detected could be very different under 'bulk' and x-ray photon spectroscopy (XPS) excitation conditions, which only examine the top $500\,\mu m$ or so. However, no author has as yet published a comparison of the SO_2–PE spectra with those of a synthesised bisulphate, e.g. structure **Ia**, but the Aberdeen group[97] has eliminated cyclic sulphates by comparison with the model compound (**III**).

Formation of dialkyl sulphate would seem to require reaction between a bisulphate of one chain and a hydroxyl in another

$$\text{HC—OSO}_2\text{OH} + \text{HO—CH} \longrightarrow \text{HC—OSO}_2\text{O—CH} + \text{H}_2\text{O}$$

On statistical grounds this seems unlikely; if it occurred to any significant extent the crosslinks so formed would be noticeable. Whereas intra-molecular chain processes, such as the following,

could certainly produce some alcohol, by no means would one expect 50 % of the hydroperoxide to be so isomerised. However, since intramolecular processes appear to have been ruled out, radical processes should be considered.

As seen in Section 5.1, various radicals are likely to be formed below the polymer surface:

$$\text{ROOH} \xrightarrow{\text{SO}_2} \text{ROSO}_2^- + \text{HOSO}_2^- + \text{RSO}_2^-$$
$$\text{ROO}^\cdot \xrightarrow{\text{SO}_2} \text{ROOSO}_2^- \longrightarrow \text{RO}^\cdot + \text{SO}_3$$

Hence

where $X = OSO_2^-$, SO_2^-, $O^{\cdot}$, which could lead to the formation of alcohol by hydrogen abstraction

$$
\begin{array}{cc}
-\mathrm{CH}- & \mathrm{C}- \\
| & \| \\
\mathrm{OH} & \mathrm{CH_2}
\end{array}
$$

or

$$
\begin{array}{cc}
-\mathrm{CH}- & \mathrm{C}- \\
| & \| \\
\mathrm{Y} & \mathrm{CH_2} \\
| & \\
-\mathrm{CH}- & \mathrm{C}- \\
& \| \\
& \mathrm{CH_2}
\end{array}
$$

where

$$
Y = \begin{cases}
\mathrm{OSO_2O} & \text{(sulphate)} \\
\mathrm{SO_2O} & \text{(sulphonate)} \\
\mathrm{OO} & \text{(peroxide)}
\end{cases}
$$

by radical recombination.

Such structures would be consistent with XPS[98,99] data. It seems possible that the sulphates identified by i.r.[97] could be formed via the alkoxy radical RO$^{\cdot}$ as the key intermediate. When a sulphated polymer is heated (100 °C/10 h), the sulphate decomposes, and pendant methylene groups completely disappear.[97] The partial loss of retained sulphur has been followed by ^{35}S radiotracer analysis[88] and is shown in Fig. 13; it suggests a two stage reaction or that two types of sulphur-containing moieties are present.

Further investigation of this reaction[97] showed that the loss of pendant methylene,[100] for oxidised samples containing previously undetectable hydroperoxide or carbonyl, is the same for all samples, regardless of the amount of sulphate originally present. Indeed, a small loss of pendant methylene when hydroperoxidised PE is treated with SO_2 at room temperature is a consequence of partial desulphation during the reaction normally carried out at c. 20 °C/18 h. The mechanism suggested[97] is as follows:

$$
\begin{array}{l}
-\mathrm{CH_2-CH-C-CH_2-} \\
\quad\quad | \quad \| \\
\quad\quad | \quad \mathrm{CH_2} \\
\quad\quad | \\
\quad\quad \mathrm{OSO_2O} \\
\quad\quad\quad | \\
-\mathrm{CH_2-CH-}
\end{array}
\longrightarrow
\begin{array}{l}
-\mathrm{CH=CH-C=CH-} \\
\quad\quad\quad\quad | \\
\quad\quad\quad\quad \mathrm{CH_3} \\
\\
+ \; \mathrm{H_2SO_4} \\
+ \; -\mathrm{CH=CH-}
\end{array}
$$

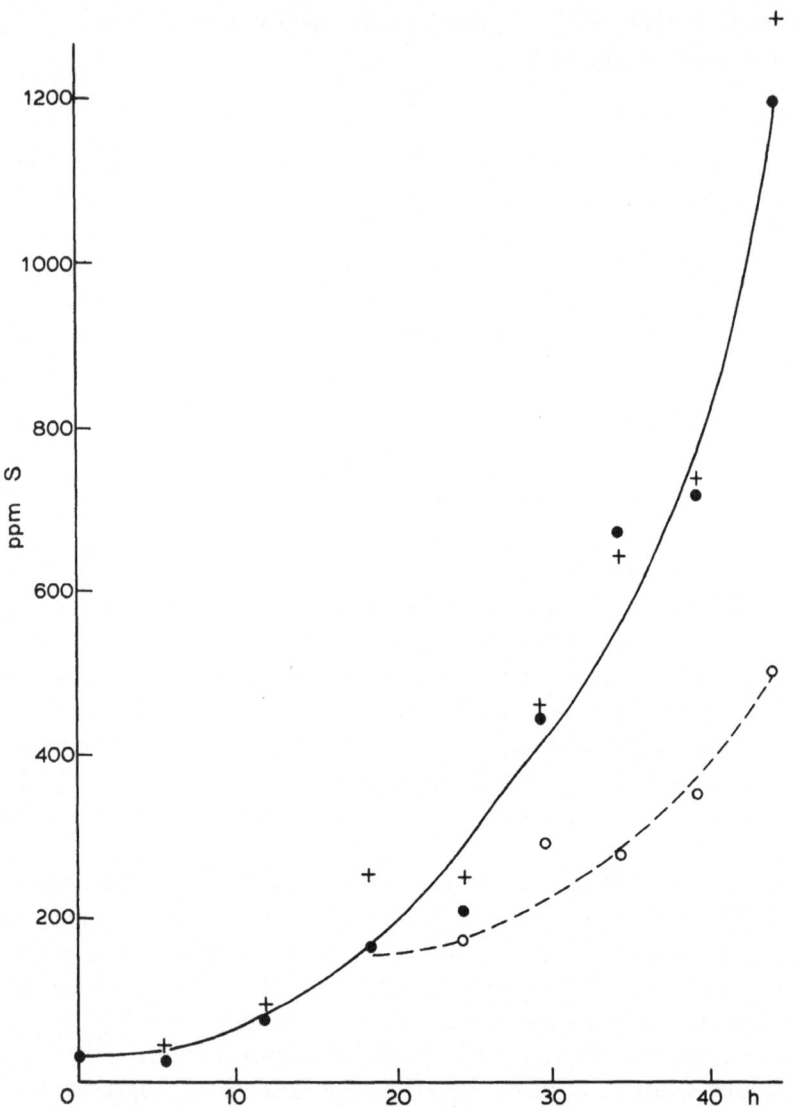

FIG. 13.   Oxidation of LDPE at 100 °C. (+) ppm OOH(I_2); (●) ppm S, as made;
(○) ppm S, after heating.

Mitchell and Perkins[39] soaked SO_2-treated pre-oxidised PE film in water
overnight and then extracted with chloroform; since i.r. spectra showed no
significant changes, they concluded that the SO_2-containing groups were
attached to the polymer chain.

A direct correlation of hydroperoxide with retained sulphur (by ^{35}S) for

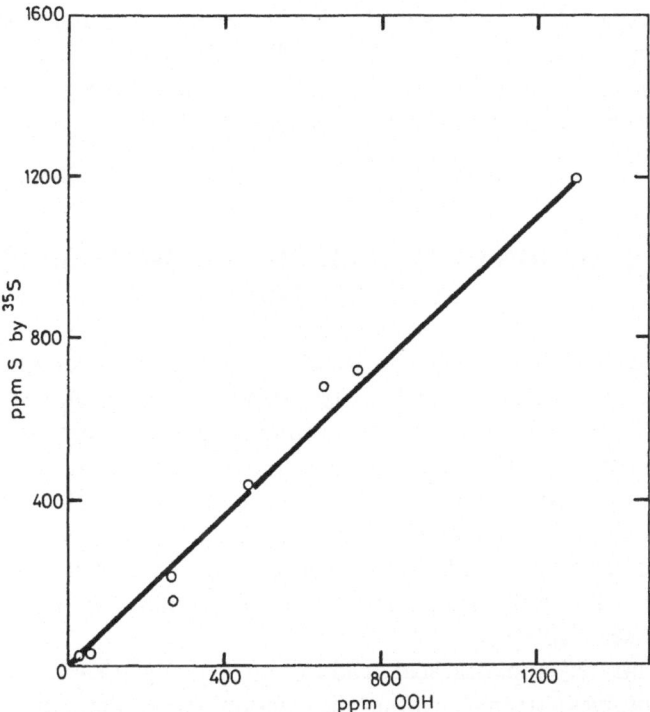

FIG. 14. Oxidised PE—correlation of OOH/S (180 μm film oxidised at 100 °C).

180 μm film oxidised at 100 °C is shown in Fig. 14. However, the ^{35}S technique is not routinely available and XRF is used; since LDPE is normally seen in the form of granules, a moulding step is often used during which sulphur losses inevitably occur. However, if the absolute sulphur values are less important than comparative differences, direct measurements can be satisfactorily obtained on unmoulded granules.[101]

5.6.3. Photo-oxidation
Unlike PP it appears[102] that the secondary hydroperoxides in PE have, surprisingly, no photo-inductive effect on the overall oxidation; in this case SO$_2$ treatment would not be expected to show any beneficial effects. However, acceleration of PE and PVC weathering due to atmospheric pollution was recorded in Japan[103] and Jellinek[104,105] has since made a special study of this topic.

5.7. HDPE
Cameron and Main[97] also looked at HDPE which had negligible pendant methylene content and found that after severe oxidation (155 °C/1·25 h)

and subsequent SO_2 treatment, changes in the i.r. spectrum similar to those for PE were recorded. The subsequent effect of heat on this treated HDPE was again similar to PE suggesting that its pendant methylene groups are lost as a consequence of desulphation and are not directly involved in the desulphation process.

6. REACTION OF POLYPROPYLENE WITH SULPHUR TRIOXIDE

When oxidised PP is treated with SO_2, a brown colour develops whose intensity increases with the hydroperoxide concentration. As sulphate was not detected in this reaction product, a possible mechanism is the formation of sulphur *tri*oxide and subsequent polymer charring. Solutions of SO_3 in chlorinated hydrocarbons react with unoxidised PP to give products physically resembling those from the SO_2 reaction, the intensity of colour increasing with SO_3 concentration. After water washing to remove residual SO_3, a high proportion of sulphur was detected in the polymer (see Table 11).

Concentrated sulphuric acid produces no charring at room temperature but oleum gives a strongly exothermic reaction. Sulphonic acid groups were directly introduced[106] into PE by treatment with fuming sulphuric acid; in similar work, PE and PP,[25,97] were treated with concentrated sulphuric acid and the sulphonic acid groups found to be thermally unstable. Reaction of both PE and PP with SO_3 is claimed[107] to reduce their permeability.

TABLE 11

REACTION OF POLYPROPYLENE WITH SULPHUR TRIOXIDE

% SO_3	Reaction solvent	% Sulphur in polymer	
		Before washing	After washing
0·1	None	0·05	0·015
0·1	CH_2Cl_2	0·08	0·03
0·2	$CHCl_3$	0·13	0·015
0·5	None	0·10	0·025
1·0	None	0·25	0·075
1·0	CH_2Cl_2	0·60	0·075
5·0	CH_2Cl_2	1·6	0·35

7. SULPHUR DIOXIDE PRODUCTION FROM SECONDARY ANTIOXIDANTS

Any paper in this area would be incomplete without reference to the fact that most sulphur-containing 'peroxide-decomposing' antioxidants have been shown[108] to operate by producing SO_2. These include the thiodipropionates,[108] dithiocarbamates,[109] dithiophosphates,[110] benzthiazoles,[110] diaryldisulphides,[111] phenolic sulphides[112,113] and thiophenes.[80,114] However, in view of Scott's work mentioned above (Section 5.4), SO_3 should perhaps now be considered as an alternative active component.

8. SULPHUR DIOXIDE AND SULPHUR HEXAFLUORIDE AS OXIDATION INHIBITORS

Even if the exact mechanisms of SO_2-hydroperoxide reactions are not clear, their possible use in extending the induction period of polyolefines can be profitably examined. In the presence of irradiation, SO_2 has long been known[115] to stabilise PE; investigation of the photochemical reaction has shown the formation of sulphones[116] and sulphinic acids.[117]

Sulphur hexafluoride is a stable insulating material used in high voltage apparatus; degradation of PE subjected to radiofrequency discharges has been studied.[118] β-Irradiation in an atmosphere of SF_6 is claimed[119] to make PE stable to outdoor weathering for 12 months; this gas is surprisingly more than twice as soluble as oxygen in amorphous PE.[120] SO_2 and SF_6 pretreatment of PE before thermal oxidation at 100 °C (Table 12) clearly reduces the amount of degradation but the fact that only 25 % reduction is obtained (cf. the protection that would be afforded by modest amounts of phenolic antioxidant) suggests that some of the

TABLE 12

EFFECT OF SO_2 AND SF_6 PRE-TREATMENT ON PE OXIDATION[a]

Pretreatment	Post-treatment	S (ppm) by XRF
SO_2	SO_2	420
SF_6	SO_2	420
None	SF_6	< 5
None	SO_2	550

[a] After 100 °C/48 h in circulating air oven.

TABLE 13

EFFECT OF SO_2 AND SF_6 PRETREATMENT ON PP OXIDATION

Hours at 100°C	S (ppm) after pretreatment			MFI 190/10 after pretreatment	
	Control	SO_2	SF_6	Control	SO_2
0	<5	10		45	
6	170	80		60	55
8·5	550	65	260	220	80
24	5000	4000			

sulphur-containing groups are labile and produce free radicals at 100 °C. Formation of dienes by loss of sulphuric acid would presumably lead to allylic oxidation.

Reaction of oxidised PE with SF_6 clearly does not lead to the formation of involatile sulphur-containing groups; there seems no obvious chemical reaction in which SF_6 can react with potential radical-producing species, and the presence of low concentrations of more reactive impurities (e.g. F_5SOSF_5) has been suggested.[12]

The effects of SO_2 and SF_6 on PP are shown in Table 13; both have an inhibiting effect. The reduction in oxidation has been examined further (Scheme 4); again it is clear that SF_6 is decomposing some of the groups that would otherwise react with SO_2.

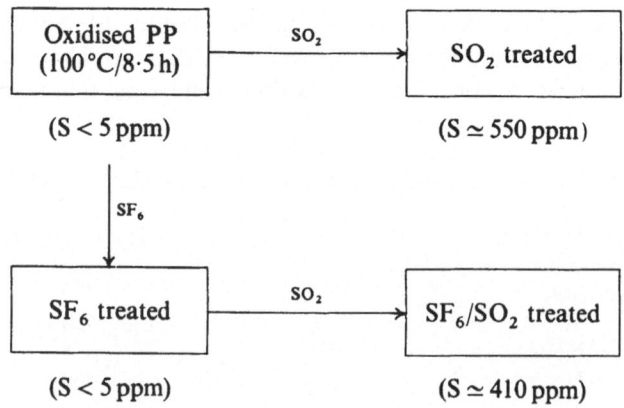

SCHEME 4. Reaction of SO_2/SF_6 with oxidised PP.

9. CONCLUSION

The reaction of SO_2 with oxidised polyolefines is the basis of new analytical tests for detecting oxidation that are several times more sensitive than previous hydroperoxide or carbonyl methods. Study of the reaction in PE confirms that hydroperoxidation probably takes place in the β-position to a pendant methylene group. The sulphur-containing groups are considerably less thermally stable in PP. Elucidation of the exact structures of sulphur-containing species and investigation of the effects of SO_2 and SF_6 on the early stages of polyolefine oxidation should prove fruitful areas for further study.

ACKNOWLEDGEMENTS

As this work began in 1969, it is not possible to record the names of all those who have contributed. However, special thanks are due to my colleagues at ICI (Mr J. Chalmers, and especially Mr D. Kingsnorth, Miss T. Kollmann, Mr S. R. Oldland, Dr H. A. Willis and Dr G. M. Wood) and to the directors of that company for permission to present the initial work.[122] I also wish to acknowledge the substantial contribution by our collaborators at Aberdeen University, Dr G. G. Cameron and Mr B. R. Main, during the period 1981–83.

REFERENCES

1. HENMAN, T. J., *Dev. Polym. Stab.*, **1** (1979), 39.
2. AL-MALAIKA, S. and SCOTT, G., In: *Degradation and Stabilisation of Polyolefines*, ed. N. S. Allen, London, Applied Science Publishers, 1983, chapters 6, 7.
3. BAWN, C. E. H. and CHAUDHRI, S. A., *Polymer*, **9** (1968), 113.
4. CHIEN, J. C. W. and WANG, D. S. T., *Macromolecules*, **8** (1975), 920.
5. BOLLAND, J. L., *Quart. Rev.*, **3** (1949), 1.
6. HENMAN, T. J., In: *Degradation and Stabilisation of Polyolefines*, ed. N. S. Allen, London, Applied Science Publishers, 1983, chapter 2.
7. IRING, M., KELEN, T., TÜDŐS, F. and LÁSZLÓ-HEDVIG, Z., *Magy Kern Foly*, **82** (1976), 244.
8. ADAMS, J. H., *J. Polym. Sci. A1*, **8** (1970), 1077.
9. CHIEN, J. C. W., VANDENBERG, E. J. and JABLONER, H., *J. Polym. Sci. A1*, **6** (1968), 381.
10. CHIEN, J. C. W. and JABLONER, H., *J. Polym. Sci. A1*, **6** (1968), 393.
11. SHILOV, Y. B. and DENISOV, Y. T., *Vys. Soed.*, **A19** (1977), 1244.

12. SHLAPNIKOV, YU. A., BOGAEVSKAYA, T. A., KIRYUSHKIN, S. G. and MONAKHOVA, T. V., *Eur. Pol. J.*, **15** (1979), 737.
13. VINK, P., In: *Degradation and Stabilisation of Polyolefines*, ed. N. S. Allen, London, Applied Science Publishers, 1983, chapter 5.
14. ADAMS, J. H., *J. Polym. Sci. A1*, **8** (1970), 1279.
15. REICH, L. and STIVALA, S. S., *J. Appl. Polym. Sci.*, **13** (1969), 17.
16. TATARENKO, L. A. and PŮDOV, V. S., *Vys. Soed.*, **10B** (1968), 287.
17. REICH, L. and STIVALA, S. S., *Autoxidation of Hydrocarbons and Polyolefines*, New York, Marcel Dekker, 1969.
18. RICHTERS, P., *Macromolecules*, **3** (1970), 262.
19. CHIEN, J. C. W., *J. Polym. Sci. A1*, **6** (1968), 381.
20. DENISOV, E. T., *Dev. Polym. Stab.*, **5** (1982), 23.
21. ADAMS, J. H., *J. Polym. Sci. A1*, **8** (1970), 1077.
22. IRING, M., KELEN, T., TÜDŐS, F. and LÁSZLÓ-HEDVIG, Z., *I. Polym. Sci. Symposia*, **57** (1976), 89.
23. AMIN, M. U., SCOTT, G. and TILLEKERATNE, L. K. M., *Eur. Polym. J.*, **11** (1975), 85.
24. CHALMERS, J. and WILLIS, H. A., unpublished work, 1982.
25. CAMERON, G. G. and MAIN, B. R., *Polym. Deg. Stab.*, **5** (1983), 215.
26. CAGLAO, M. E., RUEDA, D. R. and BALTA-CALLEJA, F. J., *Polym. Bull.*, **3** (1980), 305.
27. GOLDFARB, L., FOLTZ, C. R. and MESSERSMITH, D. C., *J. Polym. Sci., Polym. Chem. Ed.*, **10** (1972), 3289.
28. GUILLET, J. E., *Sixth International Conference on Advances in Stabilisation and Controlled Degradation of Polymers*, Luzern, 1984.
29. PERENYI, K., *Muan es Gumi*, **21** (1984), 20.
30. SOMERSALL, A. C. and GUILLET, J. E., *Polymer Preprints*, **25** (1984), 296.
31. STIVALA, S. S., KIMURA, J. and GABBAY, J. M., In: *Degradation and Stabilisation of Polyolefines*, ed. N. S. Allen, London, Applied Science Publishers, 1983, chapter 3.
32. ALLEN, N. S., In: *Degradation and Stabilisation of Polyolefines*, ed. N. S. Allen, London, Applied Science Publishers, 1983, chapter 8.
33. HENMAN, T. J. and OLDLAND, S. R., In: *Developments in Polymer Stabilisation—7*, ed. G. Scott, London, Elsevier Applied Science Publishers, 1984, chapter 6.
34. FONGEA, D., GHALEB, M., GERARD, P. and PINERI, M., *Rev. G. Caout. Plast.*, **49** (1972), 1063.
35. BILLINGHAM, N. C., personal communication, 1981.
36. GANITSKII, A. B., GANITSKII, M. B., IVANOV, A. P., IVANOV, R. A., KAZIMIROV, L. M. and PIOTROVSKII, K. B., *Kauch. i Rezina*, **27** (1968), 44.
37. HANSON, D. E., ICI Technical Service Note PP132, 1978.
38. BURNETT, J. D., MILLER, R. G. J. and WILLIS, H. A., *J. Polym. Sci.*, **15** (1955), 594.
39. MITCHELL, J. and PERKINS, L. R., *Appl. Polym. Symposia*, **4** (1967), 167.
40. HENMAN, T. J., *World Index of Polyolefine Stabilisers*, London, Kogan Page/Royal Society of Chemistry, 1982.
41. ALLEN, N. S., FATINIKUM, K. O. and HENMAN, T. J., *Polym. Deg. Stab.*, **4** (1982), 59.

42. BILLINGHAM, N. C., *Fourth International Conference on Advances in Stabilisation and Controlled Degradation of Polymers*, Luzern, 1982.
43. GETMANENKO, E. N. and PEREPLATCHIKOVA, E. M., *Zh. Anal. Khim.*, **29** (1974), 830.
44. KINGSNORTH, D. J. and WOOD, D. G. M., unpublished work, 1978.
45. BOCEK, P., *Chem. Prum.*, **17** (1967), 439.
46. JADRNICEK, B., *Chem. Prum.*, **15** (1965), 681.
47. POKORNY, B., *Chem. Prum.*, **18** (1968), 315.
48. Chemie-Linz AG, personal communication, 1979.
49. ALLEN, N. S., personal communication, 1982.
50. HAWKINS, W. L., *Polym. News*, **4** (1978), 279.
51. HAWKINS, W. L., *Polymer Degradation and Stabilisation*, Berlin, Springer-Verlag, 1984, p. 100.
52. HORNG, P. L. and KLEMCHUK, P. L., *Polymer Preprints*, **25** (1984), 76.
53. MALTESE, P., *Mater. Plast. Elast.*, **36** (1970), 69.
54. TSUJI, K., *ACS Div. Org. Coatings Plast. Chem. Pap.*, **35** (1975), 167.
55. WINDLE, J. J. and FREEDMAN, B., *J. Appl. Polym. Sci.*, **21** (1977), 2225.
56. KHOLMOGROV, V. E., *Russ. Chem. Rev.*, **37** (1968), 628.
57. KLINSHPONT, E. R., MILINCHUK, V. K., PASHCHENKO, V. I. and GILYAZITDINOV, D. G., *Vys. Soed.*, **16A** (1974), 49.
58. GULRAJANI, M. L., *Silk Tayon Ind. India*, **13** (1970), 263; *Chem. Abs.*, **73**, 121099.
59. SVOBODA, P., *Anal. Fys. Metody Vyzk Plastu. Pryskyric*, **1** (1971), 230; *Chem. Abs.*, **76**, 4399.
60. PEELING, J. and CLARK, D. T., *J. Polym. Sci. Polym. Chem.*, **21** (1983), 2047.
61. CLOUGH, R. and GILLEN, K., *Polymer Preprints*, **25** (1984), 83.
62. ONISCHUK, V. A., PŮDOV, V. S., SHANTAROVICH, V. P. and YASINA, L. L., *Polym. Sci. USSR*, **24** (1982), 2966.
63. BOUSTEAD, I., *Proc. Roy. Soc. Lond. A*, **318** (1970), 459.
64. ALLEN, N. S., In: *Analysis of Polymer Systems*, eds. L. S. Bark and N. S. Allen, London, Applied Science Publishers, 1982, p. 79.
65. GEORGE, G. A., *Dev. Polym. Degrad.*, **3** (1981), 173.
66. ZLATKEVICH, L., *Abstracts 187th ACS Meeting*, St Louis, April 1984; *Polymer Preprints*, **25** (1984), 81.
67. NATHAN, R. A., MENDENHALL, G. D. and HASSELL, J. A., *Proc. Polym. Charac. Conf.*, **1** (1976), 123.
68. DAVENPORT, J. E. and MAYO, F. R., *Polymer Preprints*, **25** (1984), 79.
69. MENDENHALL, G. D., *Angew. Chem.*, **89** (1973), 220.
70. WENDLANDT, W. W., *Therm. Acta.*, **72** (1984), 363.
71. MONACO, S. B. and RICHARDSON, J. H., *Polym. News*, **9** (1984), 230.
72. MATISOVA-RYCHLA, L., RYCHLY, J. and VAVREKOVA, M., *Eur. Polym. J.*, **14** (1978), 1033.
73. BILLINGHAM, N. C., personal communication, 1983.
74. REICH, L. and STIVALA, S. S., *Makromol. Chem.*, **103** (1967), 74.
75. IRING, M., LÁSZLÓ-HEDVIG, Z., BARABÁS, K., KELEN, T. and TÜDŐS, F., *Eur. Polym. J.*, **14** (1978), 439.
76. FORTUIN, J. P. and WATERMAN, H. I., *Chem. Eng. Sci.*, **2** (1954), 182.

77. MAZZOLINI, C., PATRON, L., MORETTI, A. and CAMPANELLI, M., *Ind. Eng. Chem. Prod. Res. Develop.*, **9** (1970), 504.
78. FLOCKHART, B. D., IVIN, K. J., PINK, R. C. and SHARMA, B. D., *J. Chem. Soc. D. Chem. Commun.* (1971), 339.
79. GILBERT, B. C., NORMAN, R. O. C. and SEALY, R. G., *J. Chem. Soc. Perkin II* (1975), 308.
80. VAN TILBORG, W. J. M. and SMAEL, P., *Rec. Trav. Chim.*, **95** (1976), 138.
81. HUSBANDS, M. J. and SCOTT, G., *Eur. Polym. J.*, **15** (1979), 249.
82. TOCHIN, V. A., SHYAKHOV, R. A. and SAPOZHNIKOV, D. N., *Polym. Sci. USSR*, **22** (1980), 830.
83. DAVIS, E. G. and ROONEY, M. L., *Koll.-Z. Z. Polym.*, **249** (1971), 1043.
84. BOSS, C. R. and CHIEN, J. C. W., *J. Polym. Sci. A1*, **4** (1966), 1543.
85. KIRYUSHKIN, S. G. and GROMOV, B. A., *Vys. Soed.*, **14A**(viii) (1972), 1715.
86. FELDER, R. M., SPENCE, R. D. and FERRELL, J. K., *J. Chem. Eng. Data*, **20**(iii) (1975), 235.
87. HENMAN, T. J., unpublished work, 1969.
88. MAIN, B. R. and OLDLAND, S. R., unpublished work, 1983.
89. ALLEN, N. S., *Polym. Deg. Stab.*, **2** (1980), 155.
90. CARLSSON, D. J., personal communication, 1984.
91. WOOD, D. L. and LUONGO, J. P., *Mod. Plast.* (1961), 132.
92. ALLEN, N. S., FATINIKUN, K. O. and HENMAN, T. J., *Chem. Ind.* (1981), 119.
93. FARMER, E. H., *J. Chem. Soc.* (1942), 340.
94. SADRMOHAGHEGH, C. and SCOTT, G., *Eur. Polym. J.*, **16** (1980), 1037.
95. MITCHELL, J. and SMITH, D. M., *Aquametry*, New York, Interscience, 1948, p. 142.
96. OLDHAM, W. J. and WIRTH, M. M., UK Patent 734 403, 1950.
97. CAMERON, G. G. and MAIN, B. R., unpublished work, 1982.
98. BRIGGS, D. and KENDALL, C. R., *Int. J. Adhes. Adhes.*, **199** (1982), 13.
99. MUNRO, H., personal communication, 1983.
100. CHALMERS, J., HENMAN, T. J. and WILLIS, H. A., unpublished work, 1982.
101. REDFERN, D., personal communication, 1983.
102. GINHAE, J.-M., GARDETTE, J.-L., ARNAUD, R. and LEMAIRE, J., *Makromol. Chem.*, **182** (1981), 1017.
103. KUBOTA, H., NISHIMURA, O. and SUZUKI, S., *Prog. Jap. Congr. Mater. Res.*, **24** (1981), 282; *Chem. Abs.*, **95**, 188017.
104. JELLINEK, H. H. G., *Text. Res. J.*, **43** (1973), 557.
105. JELLINEK, H. H. G., FLAJSMAN, F. F. and KRYMAN, F. J., *J. Appl. Polym. Sci.*, **13** (1969), 107.
106. OLSEN, D. A. and OSTERAAS, A. J., *J. Polym. Sci. A1*, **7** (1969), 1921.
107. Sulzer Brothers, Dutch Patent 74/09578, 1974.
108. ARMSTRONG, C., HUSBANDS, M. J. and SCOTT, G., *Eur. Polym. J.*, **15** (1979), 241.
109. CHAKRABORTY, K. B. and SCOTT, G., *Eur. Polym. J.*, **13** (1977), 1007.
110. ARMSTRONG, C., INGHAM, F. A. A., PIMBLOTT, J. G., SCOTT, G. and STUCKEY, J. E., *Int. Rubb. Conf.*, Brighton, 1972, F2.1.
111. HAWKINS, W. L. and SATTER, H., *J. Polym. Sci. A1* (1963), 3499.
112. CHASAR, D. W., *Polym. Deg. Stab.*, **3** (1980–81), 121.

113. BRIDGEWATER, A. J. and SEXTON, M. D., *J. Chem. Soc. Perkin II* (1978), 530.
114. VAN TILBORG, W. J. M. and SMAEL, P., *Rec. Trav. Chim.*, **95** (1976), 132.
115. GRACE, W. R., US Patent 3 464 952, 1969.
116. DEGTYAREVA, A. A., KACHAN, A. A., SHAROVOL'SKAYA, L. N. and SHRUBOVICH, V. A., *Polym. Sci. USSR*, **17** (1975), 2471.
117. DEGTYAREVA, A. A., SHAROVOL'SKAYA, L. N., SHRUBOVICH, V. A. and KACHAN, A. A., *Sint Fiz Khim. Polim.*, **15** (1975), 147; *Chem. Abs.*, **83**, 165004.
118. GILBERT, R., CASTONGUAY, J. and THEORET, A., *J. Appl. Polym. Sci.*, **24** (1979), 125.
119. Matsushita Electric. Japanese Patent 73 24 826, 1973.
120. MICHAELS, A. S. and BIXLER, H. J., *J. Polym. Sci.*, **50** (1961), 393.
121. ROBERTS, H., personal communication, 1982.
122. Presented in part at *Fifth International Conference on Advances in Stabilisation and Controlled Degradation of Polymers*, Zurich, 1983.
123. ALLEN, N. S., FATINIKUN, K. O., and HENMAN, T. J., *Eur. Polym. J.*, **19** (1983), 551.

Chapter 5

DEGRADATION OF VIRGIN AND MODIFIED CHLORINE CONTAINING POLYMERS

F. Tüdős, B. Iván, T. Kelen

Central Research Institute for Chemistry, Hungarian Academy of Sciences, Budapest, Hungary

and

J. P. Kennedy

Institute of Polymer Science, University of Akron, Ohio, USA

SUMMARY

Thermal and thermooxidative degradation of labile tertiary and/or allylic chlorine containing polymers, PVC, polychloroprene (CR), chlorinated butyl rubber (Cl-IIR), chlorinated ethylene–propylene copolymers (Cl-EPM) and polyisobutylenes (PIB) carrying tertiary chlorine chain ends have been studied.

PVCs possessing increased concentrations of internal allylic chlorines, PVC(A)s, have been prepared by controlled chemical dehydrochlorination. Tertiary chlorine containing PVCs have been synthesised by copolymerisation of vinyl chloride with 2-chloropropene. Both allylic and tertiary chlorines significantly decrease the thermal and thermooxidative stability of the resin. Systematic thermal degradation studies using a number of techniques (HCl loss, ultraviolet and visible spectroscopy, ozonolysis, gel permeation chromatography measurements) in inert solution prove quantitatively that random initiation of HCl loss at normal —CH_2—CHCl— repeat units also takes place. The rate constant of this process is lower by four orders of magnitude than the rate constant for initiation of HCl loss at internal allylic chlorines.

147

Results obtained with allylic or tertiary chlorine containing elastomers, i.e. CR, Cl-IIR, Cl-EPM and PIBs containing tertiary chlorine chain ends, also indicate the extremely low thermal stability of these labile sites.

Cationic modifications, e.g. cyclopentadienylation by |Me₃CpAl, Me₂Al treatment and grafting with isobutylene, significantly increase the thermal stability of PVC(A)s, PVC(T)s and the labile halogen containing rubbers. This is due to the replacement of tertiary and allylic chlorines by more stable groups, i.e. cyclopentadienyl, methyl or grafted PIB branches. Combining thermal dehydrochlorination and Me₃Al treatment of labile chlorine containing elastomers has led to a new analytical method for the determination of the concentrations of tertiary and/or allylic chlorines in these polymers.

Grafting with isobutylene increased the thermooxidative stability of PVC(A)s and PVC(T)s, while cyclopentadienylation has resulted in lower thermooxidative stability. This is attributed to the high oxidisability of cyclopentadienyl (Cp) groups. The oxidation of Cp groups may yield peroxy radicals which are able to initiate subsequent HCl loss by attacking normal repeating units in PVC.

1.   INTRODUCTION

Chlorine containing polymers, such as PVC, polychloroprene (CR), chlorinated butyl rubber (Cl-IIR), chlorinated polyethylene (Cl-PE), chlorinated ethylene–propylene copolymers (Cl-EPM), etc., are widely used in a variety of applications on account of their attractive physical–mechanical–chemical properties. Unfortunately, most of these materials exhibit rather poor thermal and/or thermooxidative stabilities.[1-6] Thus, sustained efforts have been made by academic and industrial research groups to elucidate the degradation behaviour and stabilisation of these polymers. However, in spite of intensive worldwide research, certain essential details of the degradation mechanism and kinetics still remain obscure.

Among the chlorine containing polymers, PVC is undoubtedly the most important in terms of tonnage produced. It is generally accepted that PVC loses HCl in a series of allylic chlorine activated steps (zipping) accompanied by increasing discolouration due to the formation of sequences of conjugated double bonds (polyenes). The physical properties of the resin may change drastically leading to objectionable products even at conventional processing temperatures. The fundamental problem is to correlate the structure with the stability of the polymer.

After the *2nd International Symposium on PVC*, held in 1976, in Lyon,

France, a number of workers became interested in the identification of structural irregularities of PVC believed to be responsible for the reduced stability of the resin (for reviews see, e.g., refs 1, 2, 5–7, 9, 10). An international Working Party on PVC Defects has also been organised under the auspices of IUPAC. The main purpose of this international cooperation is to generate quantitative information as to the nature and concentration of structural irregularities, and their relationship to the stability of PVC. Since the concentration of irregularities and defects is extremely low in commercial PVCs their identification and quantification are very difficult tasks. The examination of the detailed structures and degradation behaviours of various PVCs would be highly desirable.

Considerable research has been carried out, and will be examined concerning the synthesis of PVCs containing readily detectable concentrations of 'labile', i.e. allylic, and/or tertiary chlorine. Systematic studies of the degradation behaviour of these polymers could lead to valuable quantitative results relative to the structure–stability relationship. Surprisingly very few papers were published on this subject during the 1970s[11–13] but research has recently become more intense[5,14–27]

The degradation of other chlorine containing polymers, possessing isolated allylic or tertiary chlorines, e.g. with tertiary chlorine chain ends,[28] virgin and cationically modified[3,29–31] Cl-IIR[3] and Cl-EPM[3] has also been studied.

In spite of the practical importance of PVC degradation in the presence of oxygen, much less attention has been paid to the thermooxidative degradation than to the thermal degradation and stabilisation of PVC. Although many details of the thermooxidative degradation of PVC are still unknown, some of the most characteristic features are relatively well understood.[4] Thus one of the main problems which is still unsolved is the role of oxygen-sensitive irregularities in PVC in the thermooxidative degradation of the resin. It is expected that thermooxidative degradation studies with labile halogen-enriched and/or chemically modified PVCs would yield valuable insight in this direction.

Tertiary and allylic chlorines are also active in cationic reactions. Kennedy and his co-workers have demonstrated that tertiary and allylic chlorine containing compounds are effective cationic initiators[32–34] in conjunction with suitable co-initiators, such as certain alkylaluminiums (Me_3Al, Et_3Al, Et_2AlCl etc.). This has made possible the cationic modification of labile halogen containing polymers.[3,19,21–24,29,31,35–37] PVCs with enhanced thermal stability have been obtained on treatment with alkylaluminiums[36] and cationic grafting.[35] Also, a wide variety of novel materials, i.e. graft copolymers,[35] cyclopentadienylated rubbers (thermally reversible networks),[37] have been synthesised by cationic derivatisation.

This chapter summarises our recent results on the synthesis of allylic chlorine enriched PVCs (PVC(A)) by controlled chemical dehydrochlorination, the synthesis of PVCs containing an enhanced amount of tertiary chlorine (PCT(T)) and the degradation of these and/or modified PVCs and of some additional labile chlorine containing elastomers. Our own work and that of others will be critically examined in the light of recent results.

2. SYNTHESIS OF ALLYLIC AND TERTIARY CHLORINE ENRICHED PVCs

2.1. Introduction of Allylic Chlorines into PVC by Controlled Chemical Dehydrochlorination

Great attention has recently been focused on the effect of labile sites in PVC relative to the degradation behaviour of the resin. Among these defects allylic chlorines seem to be the most labile. However, the concentration of these chlorines is extremely low in commercial products which makes it very difficult to determine the real effect of these defects on the degradation behaviour of PVC. Moreover, strong intercorrelation may also exist between parallel processes in PVC degradation, e.g. initiation of HCl loss at defects other than allylic chlorines, random initiation, HCl-catalysis, etc. Thus, we became interested in the preparation of PVCs containing increased and therefore readily detectable amounts of internal (in-chain) allylic chlorines.

In the absence of side reactions, such as substitution, dehydrochlorination of PVC by a strong base should exclusively yield in-chain polyene sequences terminated by allylic chlorines. To avoid side reactions and to generate only allylic chlorines, potassium tert.-butoxide (t-BuOK), a tetrahydrofuran (THF) soluble relatively poorly nucleophilic strong base, has been selected as dehydrochlorinating agent:[19,20]

$$PVC\text{---}(CH\text{---}CH_2)_n\text{---}CH\text{---}CH_2\text{---}$$
$$\qquad\qquad | \qquad\qquad\quad |$$
$$\qquad\qquad Cl \qquad\qquad\;\; Cl$$

$$-nHCl \;\Big|\; THF/t\text{-BuOK}$$
$$\downarrow$$

$$PVC\text{---}(CH\text{=}CH)_n\text{---}CH\text{---}CH_2\text{---}$$
$$\qquad\qquad\qquad\qquad\qquad |$$
$$\qquad\qquad\qquad\qquad\qquad Cl$$

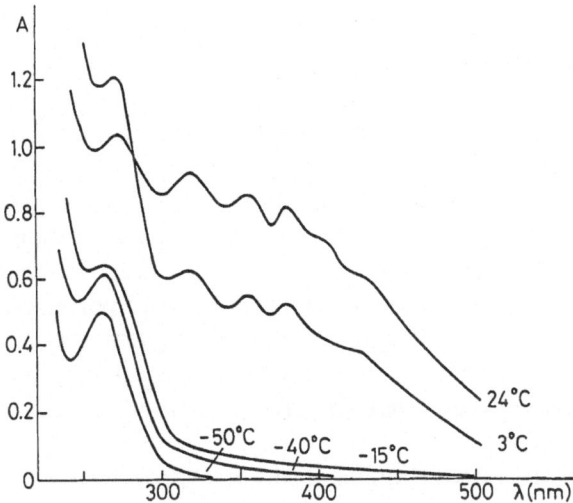

FIG. 1.   Ultraviolet–visible spectra of PVCs (0·25 % in THF) treated with t-BuOK (1 t-BuOK/100 VC) at different temperatures for 30 min.

The data in Fig. 1 and Table 1 show that dehydrochlorination of PVC by t-BuOK yields less coloured or colourless resins with correspondingly shorter polyene concentrations and lengths the lower the temperature. The fact that colourless products may be obtained in the temperature range $-30-$$-50\,°C$ indicates that dehydrochlorination under these conditions

TABLE 1

EFFECT OF DEHYDROCHLORINATION TEMPERATURE ON MOLECULAR WEIGHT, POLY-DISPERSITIES AND CHAIN SCISSION BY OZONISATION OF PVC(A) (t-BuOK/VC = 0·01) (30 min)

Temperature ($°C$)	Molecular weight and polydispersity				$S = \dfrac{\bar{M}_n^o}{\bar{M}_n} - 1$	Colour
	After dehydrochlorination		After ozonisation			
	$\bar{M}_n^o \times 10^{-3}$	$\bar{M}_w^o/\bar{M}_n^o$	$\bar{M}_n \times 10^{-3}$	$\bar{M}_w/\bar{M}_n$		
Control	64·2	1·77	57·8	1·80	0·11	Colourless
24	64·4	1·78	12·9	1·83	3·99	Brick red
3	61·2	1·81	13·1	1·78	3·70	Brownish
−15	62·8	1·80	20·4	1·76	2·07	Yellowish
−40	63·6	1·77	21·4	1·81	1·97	Slightly yellowish
−50	64·1	1·77	24·7	1·79	1·59	Colourless

leads to polyenes with less than three conjugated double bonds since polyenes containing three conjugated double bonds absorb at 286 nm.[39]

Table 1 also shows number-average molecular weights $(\bar{M}_n)$ and polydispersities $(\bar{M}_w/\bar{M}_n)$ (determined by gel permeation chromatography (GPC)) of t-BuOK-treated PVC samples, and those of the same samples after ozonisation. Ozonolysis of PVC coupled with molecular mass determination yields information about the concentration of single or conjugated double bonds in the chain.[38] The average number of scissions per polymer molecule (S), i.e. the concentration of polyene sequences due to t-BuOK treatment, decreases with increasing temperatures. According to the data in Table 1, the polydispersities remain unchanged on ozonolysis. This indicates that dehydrochlorination of PVC by t-BuOK is a random process.

In spite of the interest in chemically dehydrochlorinated PVCs (for a review see, e.g. ref. 40), very few studies[41,42] are concerned with detailed structural characterisation of such polymers. Our subsequent work with t-BuOK-treated PVCs, i.e. cationic modifications and degradation, have required PVC(A)s containing only allylic chlorines. However, undesirable substitution of allylic chlorines by OH groups occurs at higher temperatures (60 °C) during the dehydrochlorination of PVC with alcoholic KOH in THF.[42] For this reason, structural characterisation of PVC(A)s and a critical examination of some of the methods proposed for labile chlorine determination in PVC have also been carried out.

Tables 2 and 3 and Fig. 2 summarise results obtained by various methods. According to model experiments,[41] cadmium acetate ($CdAc_2$) reacts with allylic chlorines, as follows, in 60–80 % yields:

$$2R_1-CH{=}CH-\underset{\underset{Cl}{|}}{CH}-R_2 + (CH_3COO)_2Cd \longrightarrow$$

$$2R_1-CH{=}CH-\underset{\underset{\underset{\underset{CH_3}{|}}{O{=}C}}{\underset{|}{O}}}{CH}-R_2 + CdCl_2$$

The appearance of the new band at 1740 cm^{-1} (Fig. 2) associated with pendant acetate groups indicates the presence of allylic chlorines in t-BuOK-treated PVC. The concentration of polyenes determined by ozonolysis and that of allylic chlorines determined by phenolysis are essentially identical, as shown by the data in Table 2. Since phenolysis has been shown

TABLE 2

CONCENTRATION (PER 100 MONOMER UNITS) OF INTERNAL POLYENES (S) OBTAINED BY OZONOLYSIS AND LABILE CHLORINES DETERMINED BY VARIOUS TECHNIQUES IN CHEMICALLY DEHYDROCHLORINATED PVCs ($[Cl_a]$ IS THE DIFFERENCE BETWEEN LABILE Cl OBTAINED BY PHENOLYSIS OF PVC(A) AND CONTROL SAMPLES)

Sample	Internal polyenes (S) (mol %)	Labile Cl by Phenolysis (mol %)	$[Cl_a]$ (mol %)	Cl in $CdCl_2$ (mol %)	Labile Cl by thiophenolysis (mol %)
Control	0·006 2	0·03	—	<0·01	0·10
PVC(A) (I)	0·55	0·70	0·67	0·41	—
PVC(A) (II)	0·17	0·22	0·19	0·13	0·81

to be a clean reaction which can be used for the analytical determination of labile chlorine,[43] this finding indicates that undesirable side reactions (e.g. substitution of allylic chlorines) are absent during t-BuOK treatment at low temperatures and that each polyene sequence is terminated by allylic chlorine. Perichaud et al.[44] have also dehydrochlorinated PVC with t-BuOK to high yields at 30 °C and have also concluded that substitution of allylic chlorines is absent.

The data in Table 2 also show that the amount of chloride ion in $CdCl_2$ (determined by a chloride ion selective electrode), i.e. the concentrations of labile chlorines which have reacted with $CdAc_2$, are lower than those obtained by phenolysis. This is in good agreement with the fact that the reaction of $CdAc_2$ with allylic chlorine containing model compounds is incomplete.[41]

TABLE 3

CHAIN SCISSION $(S, \text{mol}\%)$ OBTAINED BY OZONOLYSIS OF PVC(A) SAMPLES AFTER TREATMENT WITH VARIOUS REAGENTS

Reagents used for PVC treatment	S (mol %)	
	Control sample	PVC(A) (III)
Control	0·006 2	0·36
$CdAc_2$	0·007 9	0·35
Phenol	0·003 8	0·31
Thiophenol	0·005 9	0·35

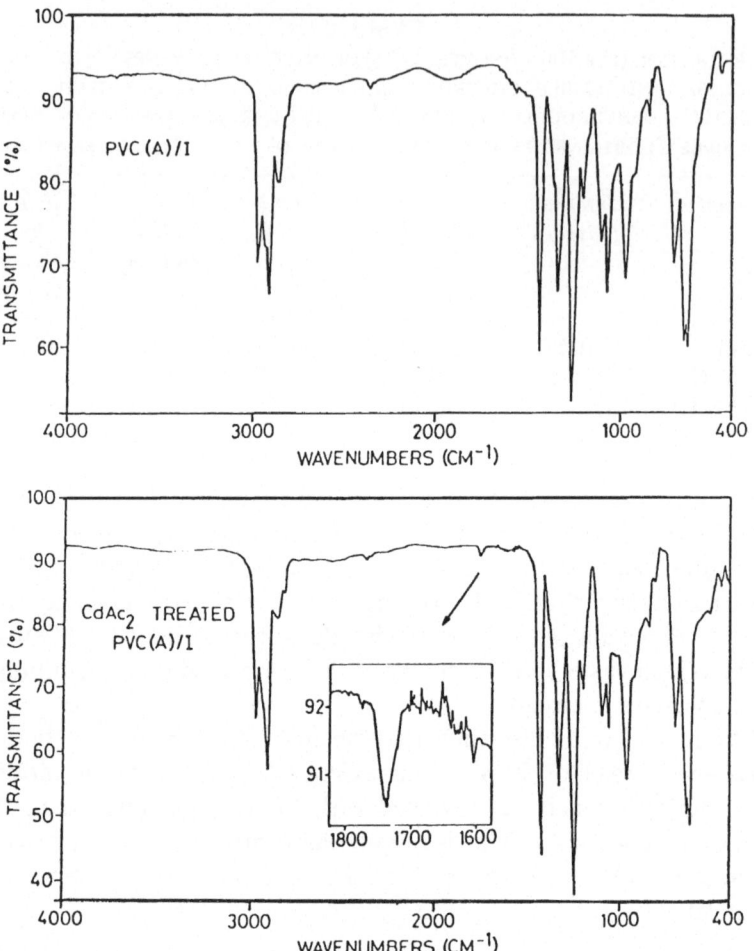

FIG. 2. Fourier transform infrared spectra of untreated and CdAc$_2$-treated PVC(A)(I) samples.

According to the data in the last column of Table 2, thiophenolysis gives significantly higher values of labile allylic chlorine than the other methods. This is rather unexpected since thiophenol has been postulated to react exclusively with allylic (and not with tertiary) chlorine:[45]

$$\text{\small\sim\sim\sim} CH{=}CH{-}CH \text{\small\sim\sim\sim} + C_6H_5{-}SH \longrightarrow$$
$$\underset{\displaystyle Cl}{|}$$

$$\text{\small\sim\sim\sim} CH{=}CH{-}CH \text{\small\sim\sim\sim} + HCl$$
$$\underset{\displaystyle \underset{\displaystyle C_6H_5}{|}}{\overset{\displaystyle S}{|}}$$

Since this method is based on the estimation of HCl evolved, the higher allylic chlorine concentrations obtained by thiophenolysis were assumed to be due to unspecified additional HCl evolution. This may be due to random HCl loss from regular repeat units. In line with this proposition, ozonolysis experiments have been carried out to examine double bond formation during thiophenolysis and other treatments (phenolysis, $CdAc_2$). According to the data shown in Table 3 the concentrations of internal polyenes in untreated and treated samples were identical within what is considered to be experimental error. Evidently, neither thiophenolysis nor the other reactions result in additional HCl elimination accompanied by double bond formation. Thus, thiophenolysis yields more HCl than there are allylic chlorines without double bond formation. Therefore, we postulate that double bond formation does not occur during dehydrochlorination by thiols. In agreement with this proposition Starnes et al.[46] have found that thiols may react not only with labile chlorines but also with the normal —CH_2—CHCl— repeating units in PVC as follows:

$$\text{—}CH_2\text{—CHCl}\text{—} + 2RSH \longrightarrow$$

$$\text{—}CH_2\text{—}CH_2\text{—} + RSSR + HCl$$

In view of this reaction, the observed discrepancy between the results of thiophenolysis and the other methods may be understood and the thiophenolysis method does not seem to be useful for the determination of allylic chlorine in PVC.

2.2. Synthesis of Tertiary Chlorine Enriched PVC by Copolymerisation of Vinyl Chloride and 2-chloropropene

PVCs containing relatively high concentrations of tertiary chlorines, PVC(T)s, have been prepared by heterogeneous free radical copolymerisation of vinyl chloride (VC) and 2-chloropropene (2-CP) (cf. Table 4). Copolymerisation kinetics have been followed by dilatometry and the time dependence of conversion is shown in Fig. 3. Final conversions obtained by dilatometry were in good agreement with those determined by gravimetry. As shown by the data in Fig. 3 and Table 4, the initial rates of copolymerisation and acceleration decrease significantly with increasing concentration of 2-CP. According to the data in Table 4 the molecular weights decrease markedly with increasing initial 2-CP concentration. Evidently 2-CP is a retarder for vinyl chloride polymerisation.

TABLE 4
POLYMERISATION CONDITIONS AND CHARACTERISATION OF VINYL CHLORIDE/2-CHLOROPROPENE COPOLYMERS

Sample	Polymerisation conditions[a]		Copolymer composition[b]	Yield (%)	$\bar{M}_n$ ($\times 10^{-3}$)	$\dfrac{\bar{M}_w}{\bar{M}_n}$
	$[AIBN] \times 10^2$ (mol dm^{-3})	$[2\text{-}CP]_0$ (mol %)	2-CP (mol %)			
1	1·34	—	—	13·7	78·4	2·01
2	1·39	0·50	0·78	12·0	73·4	1·71
3	1·31	1·12	1·42	10·0	59·4	1·87
4	1·52	2·25	3·45	9·0	53·2	1·82
5	1·47	4·80	6·46	6·7	31·5	2·05

[a] Temperature 40 °C (heterogeneous polymerisation).
[b] Determined by elemental analysis.

As Table 4 shows, a series of PVC(T)s containing $\sim 0\cdot8\text{--}6\cdot5$ mol % 2-CP units, i.e. tertiary chlorines, were obtained. The data are insufficient, however, for an accurate determination of reactivity ratios r_{VC} and r_{2-CP} because of the very narrow monomer and copolymer composition range studied.

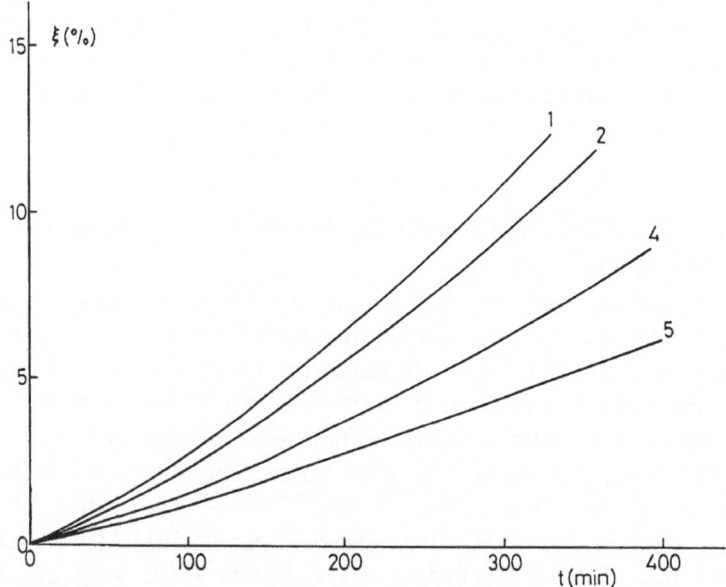

FIG. 3. Conversion as a function of time during copolymerisation of vinyl chloride and 2-chloropropene (for identification see Table 4).

3. THERMAL AND THERMOOXIDATIVE DEGRADATION OF PVC(A)

Systematic degradation studies have been carried out with a series of PVC(A) samples containing increased concentrations of allylic chlorines introduced by chemical dehydrochlorination. Table 5 shows sample numbering and allylic chlorine concentrations determined by ozonolysis coupled with molecular weight determination by GPC.

Figures 4, 5 and 6 summarise results obtained during thermal and thermooxidative degradation of solid PVC(A)s. Figures 4 and 5 show the extent of HCl loss (ξ_{HCl}) as a function of time during thermal and thermooxidative degradation, respectively. According to these data, the introduction of low concentrations of allylic chlorines into PVC significantly decreases both the thermal and thermooxidative stability of the resin. As shown by Fig. 6 a linear relationship exists between the initial rate of dehydrochlorination, $(V_{HCl})_0$, and allylic chlorine concentration. The relatively high intercept in the $(V_{HCl})_0$ versus S plot for thermal dehydrochlorination may indicate the presence of non-allylic active chlorines and/or chain defects; however, in view of the low concentration of

TABLE 5

INTERNAL ALLYLIC CHLORINE CONCENTRATION DE-
TERMINED BY OZONOLYSIS OF PVC(A) SAMPLES

Sample	Internal allylic chlorine (mol %)
PVC(G103)[a]	0·011
1	0·027
2	0·053
3	0·063
3a	0·077
PVC(S470)[b]	0·006
4	0·012
5	0·017
6	0·036
7	0·069
8	0·107

[a] Commercial suspension resin from Goodrich Chemicals.
[b] Commercial suspension resin from Borsod Chemicals.

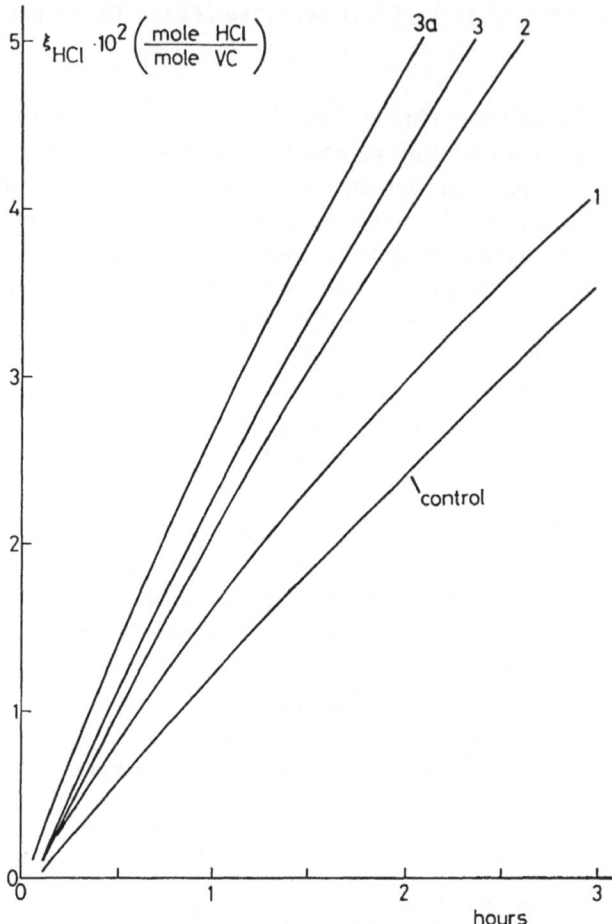

FIG. 4. Extent of HCl loss as a function of time during thermal degradation (190 °C, N_2) of PVC(A)s containing different amounts of allylic chlorines (for identification see Table 5).

these sites in PVC the intercept is most likely due to a relatively high extent of random initiation of dehydrochlorination.

If only thermal dehydrochlorination occurs during the early stages of thermooxidative degradation of PVC, then the $(V_{HCl})_0$ versus S plots for thermal and oxidative degradations would be identical. In fact, both the slope and intercept obtained for thermooxidative degradation of PVC(A)s are higher than those for thermal degradation. This may be attributed to oxidative processes operating during the early stages of thermooxidative degradation. Fast oxidation of unsaturation originally present in PVC(A)s

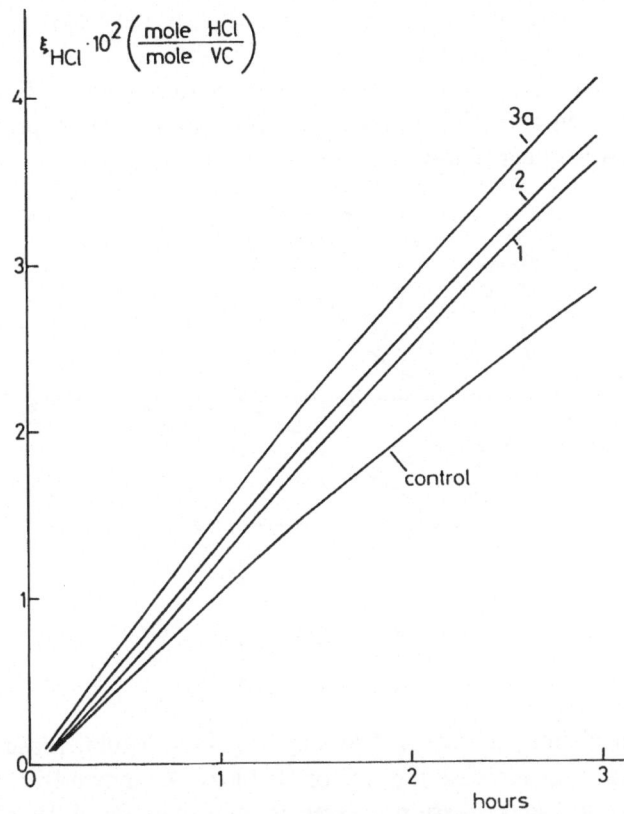

FIG. 5. Extent of HCl loss as a function of time during thermooxidative degradation of PVC(A)s (180 °C, O_2).

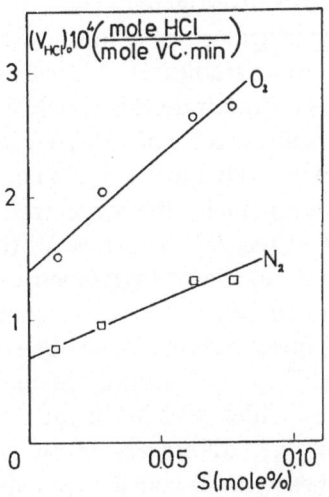

FIG. 6. Initial rates of thermal (N_2) and thermooxidative (O_2) dehydrochlorination (180 °C) as a function of concentration of internal allylic chlorines (S) of solid PVC(A)s (samples: G103, 1, 2 and 3a).

and/or formed during initial dehydrochlorination may lead to peroxy radicals according to the following scheme for the main processes of thermooxidative degradation of PVC:[4]

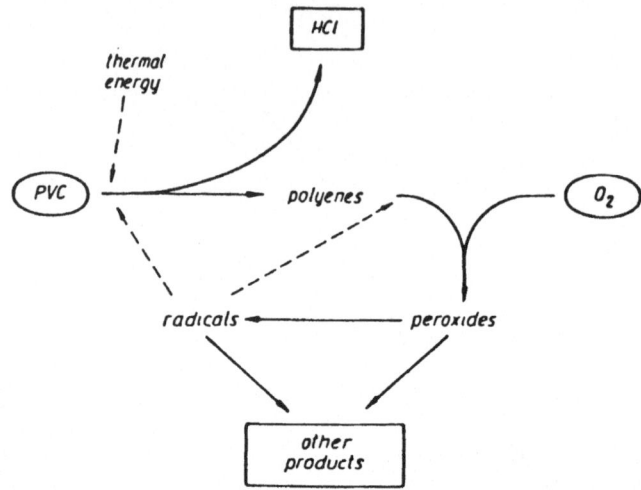

Peroxy radicals may induce HCl loss by attacking regular repeat units in PVC and will thus increase the rate of HCl loss. As shown by the $(V_{HCl})_0$ versus S plot in Fig. 6, these processes are effective even during the early stages of thermooxidative degradation since the initial rates of HCl loss in thermooxidative degradation are higher than those in thermal degradation for all the PVC(A) samples investigated. Thus, it may be concluded that double bonds containing irregularities seriously affect the thermooxidative stability of PVC. Degradation studies[21-23,47] with PVCs carrying pendant cyclopentadienyl groups corroborate this conclusion.

Figure 7 shows u.v.–visible spectra of PVC(A) (3a) and a control sample degraded to the same extent (HCl loss 0·5%) under an inert atmosphere. Absorptions in the visible region, i.e. the concentrations of longer polyenes, are higher for sample PVC(A)(3a) than those of the control. Indeed, the average length of polyenes calculated by geometric distribution[6] is 4·4 and 3·6 for PVC(A)(3a) and the untreated resin, respectively. This finding, similar to that of earlier investigations,[47] may be explained in terms of the reaction scheme[48] for HCl catalysis during thermal degradation of PVC. According to this scheme, differences in the rates of dehydrochlorination and polyene formation (i.e. differences in the rates of protonation–deprotonation of polyenes by HCl) may lead to a migration of protonated polyenes and thus to a reactivation of dehydrochlorination (zipping).

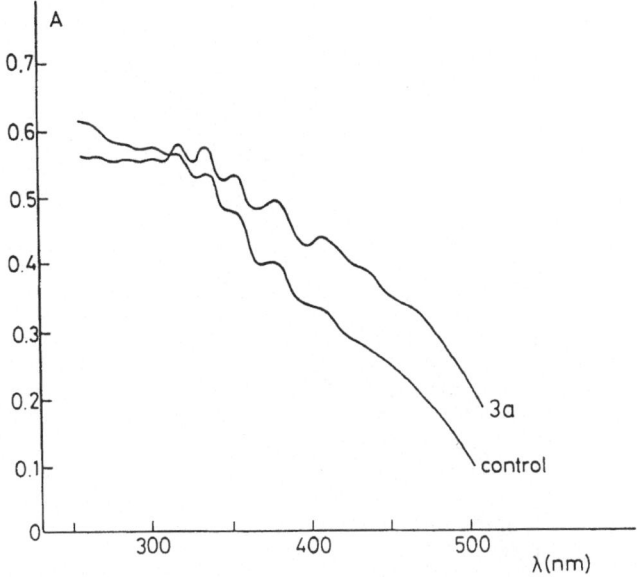

FIG. 7.   Ultraviolet–visible spectra of PVCs (G103) and PVC(A)(3a) degraded up to 0·5% HCl loss at 190 °C under N_2.

Reinitiation of HCl loss at these polyene sites yields longer polyene sequences in PVC(A)(3a), which contains higher concentrations of allylic chlorine, than in the control. Evidently, HCl may affect the thermal degradation of solid PVCs. In view of this possibility, experiments have been carried out in solution to elucidate further the effect of internal allylic chlorines on PVC degradation.

Figure 8 shows ξ_{HCl} as a function of time for PVC(A)s degraded under nitrogen in an inert solvent, 1,2,4-trichlorobenzene. According to these data, all the samples exhibit initially rapid HCl loss followed by slower rates. $(V_{HCl})_0$ increases with increasing allylic chlorine concentration, but the curves are parallel in the later stages.

Results in Fig. 9 show that while $(V_{HCl})_0$ increases linearly with allylic chlorine concentration (S), the rates in the subsequent stationary dehydrochlorination phase, $(V_{HCl})_{st}$, follow a horizontal line, i.e. the rate of HCl loss during the stationary phase is independent of the concentration of defects. These findings demonstrate the low thermal stability of internal allylic chlorines in PVC. The fact that $(V_{HCl})_{st}$ is independent of defect concentration may indicate random initiation of HCl loss at normal —CH_2—CHCl— units.

Initiation of HCl loss at units other than internal allylic chlorines may be

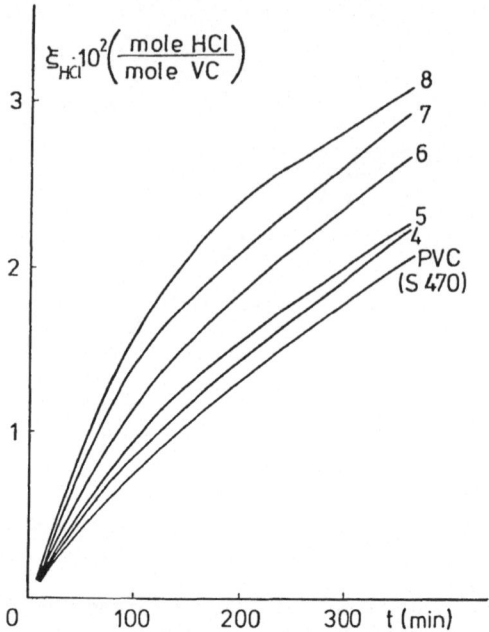

FIG. 8.   Extent of HCl loss as a function of time of PVC(A)s in TCB solution (1 %
PVC, 200 °C, N$_2$).

followed by ozonolysis combined with number-average molecular weight
determination. The concentration of internal polyene sequences (S)
initially present in the resin and formed during thermal degradation can be
obtained from

$$S = \left(\frac{1}{\bar{M}_n} - \frac{1}{\bar{M}_n^\circ} \right) M_{\text{VC}} \tag{1}$$

where $\bar{M}_n^\circ$ and $\bar{M}_n$ are the molecular weights before and after ozonisation,
respectively, and M_{VC} is the molecular weight of vinyl chloride (VC) repeat
units. As exhibited in Fig. 10, plots of concentration of internal polyene
sequences as a function of degradation times are straight lines with
identical slopes of 2.40×10^{-6}, 2.51×10^{-6} and $2.35 \times 10^{-6} \text{min}^{-1}$ for
PVC(S 470), PVC(A)(5) and PVC(A)(7), respectively. The average rate is
$2.42 \times 10^{-6} \text{min}^{-1}$. These data indicate constant rates of random initiation
of HCl loss, $(V_{\text{in}})_r$, at normal —CH$_2$—CHCl— repeat units. Since
$(V_{\text{in}})_r = k_1 c$ (where c is the concentration of —CH$_2$—CHCl— units, which
is nearly constant during the early stages of degradation) the rates of
random initiation of the control and PVC(A) samples are the same at low

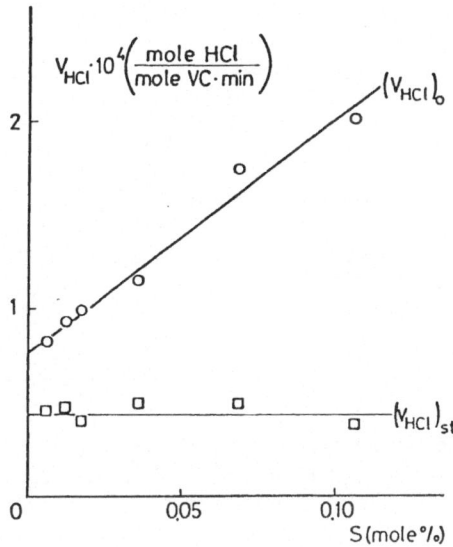

FIG. 9.   Initial, $(V_{HCl})_0$, and stationary, $(V_{HCl})_{st}$, rates of HCl loss as a function of concentration of internal allylic chlorines during thermal degradation of PVC(A)s (200 °C, N$_2$).

extents of HCl loss. The linearity of S versus time plots also indicates that the concentrations of internal defect sites other than allylic chlorines, i.e. tertiary chlorines, are negligible in the PVCs employed in this study. If this were not so, significant deviations from straight lines would be obtained in the initial parts of S versus time plots.

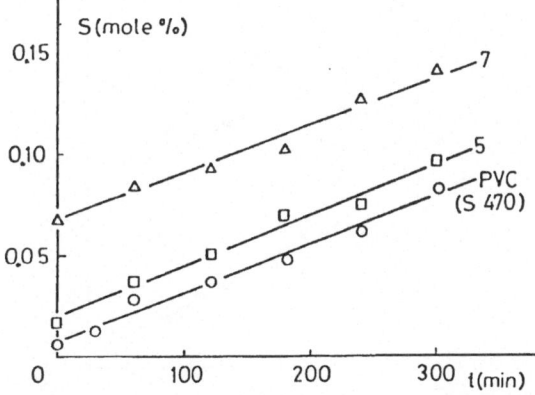

FIG. 10.   Concentrations of internal polyene sequences, determined by ozonolysis, as a function of degradation time for samples of S470, PVC(A)(5) and PVC(A)(7) during thermal degradation in TCB solution (200 °C, N$_2$).

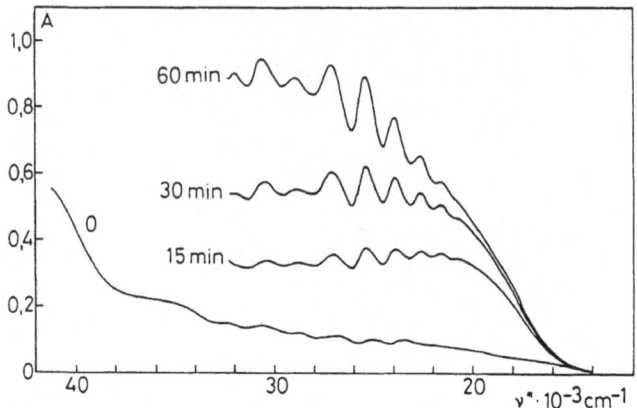

FIG. 11. Ultraviolet–visible spectra of PVC(A)(8a) (1 % PVC in THF for the undegraded sample and 0·5 % PVC in THF/TCB for the degraded ones) at various degradation times during degradation in TCB solution (N₂, 200 °C).

Ultraviolet and visible spectra of degraded PVC(A) samples have also been recorded. It was found that the concentrations of polyenes (p_i) containing different numbers of double bonds can be described by a geometric length distribution. Figure 11 shows the u.v. and visible absorbances of the original and degraded samples (8a). In contrast to samples 4–8 the undegraded 8a also contained relatively long polyenes. As

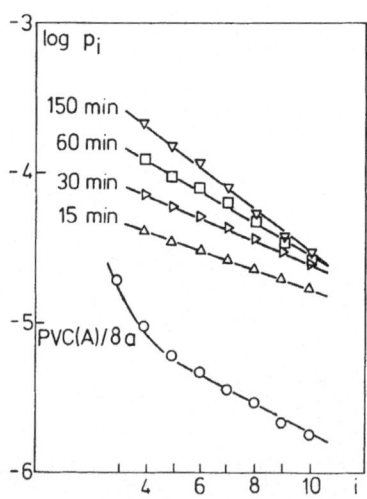

FIG. 12. Logarithm of polyene concentrations (p_i) as a function of polyene length (i) for PVC(A)(8a) at various degradation times during thermal degradation in TCB (N₂, 200 °C).

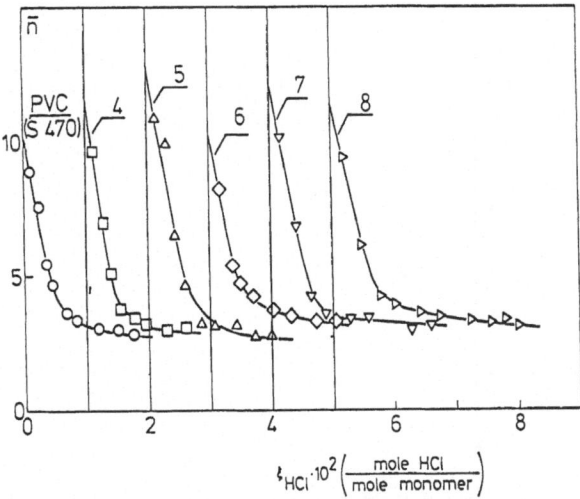

FIG. 13. Average length of polyene ($\bar{n}$) as a function of extent of HCl loss for PVC(A)s (samples 4, 5, 6, 7 and 8) during thermal degradation in TCB (200 °C, N_2); $\bar{n}$ values are shifted to the right by 1, 2, 3, 4 and 5, respectively.

shown by Fig. 12 linear log p_i versus i plots were obtained, i.e. the polyene concentrations also follow a geometric distribution with these samples at various degradation times. According to this result the type of distribution of polyenes formed during thermal degradation of PVC(A)s is independent of the distribution of polyene sequences initially present in the polymer.

Figure 13 shows the average polyene length ($\bar{n}$) as a function of HCl loss during thermal degradation of PVC(A)s. Evidently, $\bar{n}$ decreases as ξ_{HCl} increases, finally reaching a limiting constant value. Earlier it was shown that $\bar{n}$ decreases during thermal degradation of PVC and this was explained by secondary processes, i.e. cyclisation of polyenes.[49] The total concentration of labile chlorines (h_0) in PVC may be estimated using eqn (2),

$$h_0 = \frac{(\xi_{HCl})_0}{\bar{n}_0} \tag{2}$$

where $(\xi_{HCl})_0$ is the apparent 'initial' conversion determined by extrapolation of the extent of HCl loss in the stationary phase to zero time.

As shown by Fig. 14, the h_0 versus S plot is a straight line with a slope of 0·96, i.e. near unity as expected. The value of the intercept, 0·037 %, reflects the concentration of defect sites other than internal allylic chlorines in these PVCs.

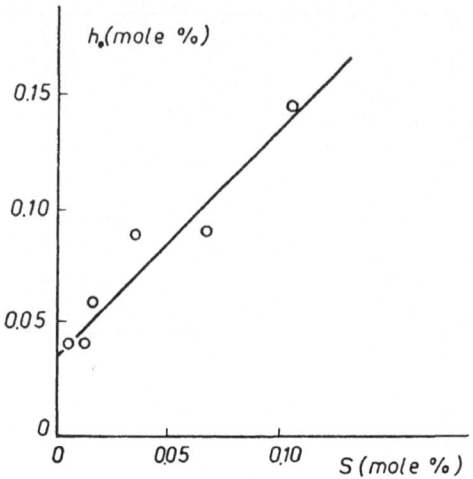

FIG. 14.   Labile chlorine concentrations (h_0) (determined by the use of eqn (2)) as a function of internal allylic chlorine concentrations (S) in PVC(A)s.

The initial rate of HCl loss at defect sites may also be calculated by the use of $\bar{n}_0$:

$$(V_{in})_0 = \frac{(V_{HCl})_0 - (V_{HCl})_{st}}{\bar{n}_0} \qquad (3)$$

According to the data in Fig. 15, $(V_{in})_0$ is a linear function of the internal allylic chlorine concentration (S). $(V_{in})_0$ may be expressed by

$$(V_{in})_0 = k^*_{1a}S + \sum_{i=1}^{N} k^*_{1i}h_i \qquad (4)$$

where k^*_{1a} is the rate constant for initiation at internal allylic chlorines and k^*_{1i} is the rate constant for initiation of HCl loss at defect sites other than internal allylic chlorines, the number of types of which is $N(=1, 2, 3, \ldots)$, and h_i is the initial concentration of such irregularities. The slope and intercept of the plot in Fig. 15 give $k^*_{1a} = 1 \cdot 06 \times 10^{-2}\,\mathrm{min}^{-1}$, and $\sum_{i=1}^{N} k^*_{1i}h_i = 2 \cdot 8 \times 10^{-6}\,\mathrm{min}^{-1}$, respectively. Dividing the latter value by the total concentration of defect sites other than internal allylic chlorines ($0 \cdot 037\%$), one obtains $0 \cdot 70 \times 10^{-2}\,\mathrm{min}^{-1}$, a rough estimate of the rate constant of initiation at such defect sites (e.g. chain end defects, tertiary chlorines). This value is only slightly smaller than k^*_{1a}, i.e. the activity and/or thermal lability of these defect structures are similar to those of the artificially introduced allylic chlorines.

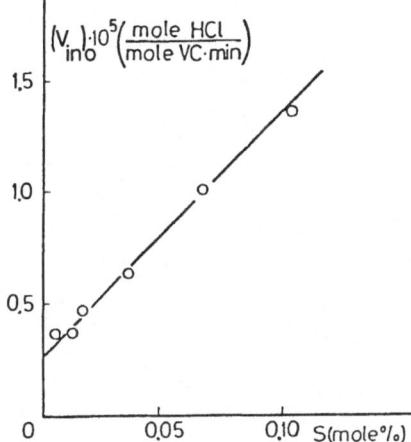

FIG. 15.   Initial rate of initiation of HCl loss at defect sites, $(V_{in})_0$, as a function of concentration of internal allylic chlorines during thermal degradation of PVC(A)s in TCB (200 °C, N_2).

From the results of ozonolysis of degraded samples shown in Fig. 10, the rate constant for random initiation of HCl loss at normal $-CH_2-CHCl-$ repeat units of PVC $k_1 = 2.42 \times 10^{-6}\,min^{-1}$. Thus, the rate constant of initiation of HCl loss at internal allylic chlorines is four orders of magnitude larger than that at normal repeat units, i.e. thermal degradation experiments with PVC(A) samples quantitatively demonstrate the low thermal stability of internal chlorines in PVC.

4.   THERMAL AND THERMOOXIDATIVE DEGRADATION OF PVC(T)

The degradation behaviour of PVC(T)s which contain relatively high concentrations of tertiary chlorine (~ 0.8–$6.5\,mol\%$), has been studied. Tertiary chlorines in PVC are believed to be thermally unstable defect sites participating in the initiation of dehydrochlorination. Figures 16 and 17 show the extent of HCl loss as a function of time for a series of PVC(T)s in nitrogen and oxygen atmospheres, respectively. According to these data, the rates of HCl loss increase with 2-chloropropene concentration in the copolymer, i.e. both thermal and thermooxidative stability decrease significantly with increasing tertiary chlorine contents (m) in PVC(T)s. This is better illustrated in Fig. 18 which shows initial rates of HCl loss,

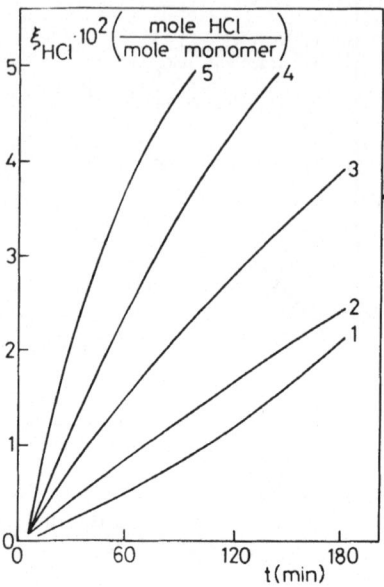

FIG. 16.   Extent of HCl loss as a function of time for PVC(T)s during thermal degradation (180 °C, N_2) (for identification of curves see Table 4).

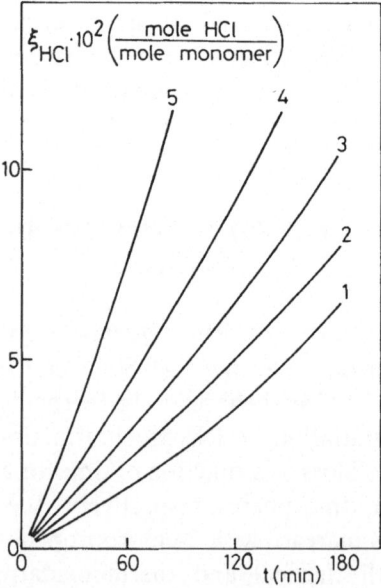

FIG. 17.   Extent of HCl loss as a function of time for PVC(T)s during thermooxidative degradation (180 °C, O_2) (for identification of curves see Table 4).

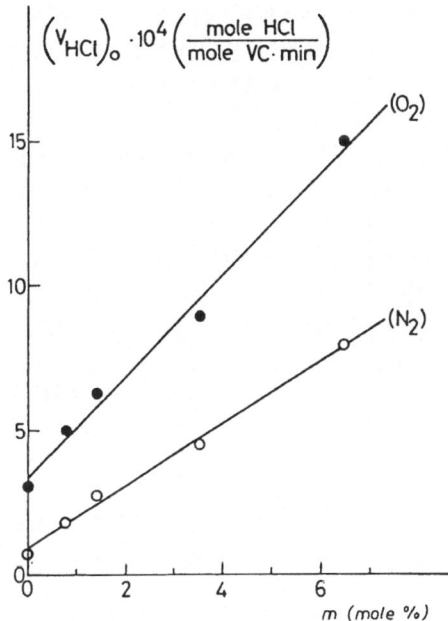

FIG. 18. Initial rates of thermal (N_2) and thermooxidative (O_2) dehydro-chlorination (180°C) as a function of concentration of tertiary chlorines in PVC(T)s.

$(V_{HCl})_0$, as a function of tertiary chlorine concentration. Evidently the rates of initial dehydrochlorination under thermal and thermooxidative degradation conditions increase linearly with the concentration of tertiary chlorines. Rates of thermooxidative dehydrochlorination are higher than those of thermal degradation, and both lines in Fig. 18 exhibit relatively significant intercepts.

According to these results tertiary chlorines in PVC significantly decrease the thermal and thermooxidative stability of the resin. The intercept in the $(V_{HCl})_0$ versus m plot, obtained for the thermal degradation of PVC(T)s may be due to the presence of irregularities other than tertiary chlorines, e.g. unsaturation and/or chain end defects, or to random initiation of HCl loss at regular monomer units. The higher rates of HCl loss during thermooxidative dehydrochlorination may be due to fast oxidation of polyenes formed during the initial stage of degradation. Peroxy radicals which emerge during this process may initiate HCl loss leading to higher rates of thermooxidative degradation.

5. THERMAL AND THERMOOXIDATIVE DEGRADATION OF CATIONICALLY CYCLOPENTADIENYLATED AND GRAFTED PVC(A)s AND PVC(T)s

Since allylic and tertiary chlorines in conjunction with certain Friedel–Crafts acids are known to be efficient sites for initiation of cationic reactions,[28-37] PVC(A)s and PVC(T)s containing relatively high concentrations of such sites were used to synthesise cyclopentadiene modified and graft copolymers. Pendant cyclopentadienyl groups (Cp) have been introduced by treatment with dimethylcyclopentadienylaluminum (Me$_2$CpAl) in PVC(A)s whose allylic chlorine concentrations were increased by mild chemical dehydrochlorination (Table 6) and in PVC(T)s synthesised by

TABLE 6

INTERNAL ALLYLIC CHLORINE CONTENTS (S) DETERMINED BY OZONOLYSIS OF PVC(A)s USED FOR CYCLOPENTADIENYLATION BY Me$_2$CpAl

Sample	$S = \dfrac{\bar{M}_n^o}{\bar{M}_n} - 1$
PVC(G103)	0·11
9	1·90
10	2·37
11	3·96

copolymerisation of vinyl chloride and 2-chloropropene (Table 4). It has recently been demonstrated[21-24,29,37] that pendant cyclopentadienyl groups in polymers lead to unique thermally-reversible networks by the Diels–Alder/*retro*-Diels–Alder condensation of Cp groups:

Although PVC proved to be a suitable starting material for cationic grafting, grafting efficiencies[35] and the number of grafted branches per PVC chain remained low because the concentrations of labile chlorines are extremely low in commercial resins.[35,50] Therefore, PVCs carrying higher concentrations of cationically active allylic and tertiary chlorines were expected to yield higher branching frequencies.

It has been shown[34,51] that allylic chlorines in conjugation with BCl_3 are excellent initiators of isobutylene polymerisation. Thus, poly(vinyl chloride-g-isobutylene)s (PVC(A)-g-PIB-Cl) have been synthesised by the use of $PVC(A)/BCl_3/i$-C_4H_8 systems. Since termination occurs by ion-collapse in these systems, PVC(A)-g-PIB-Cls with tertiary chlorine branch termini have been obtained. The following scheme helps one to visualise this process:

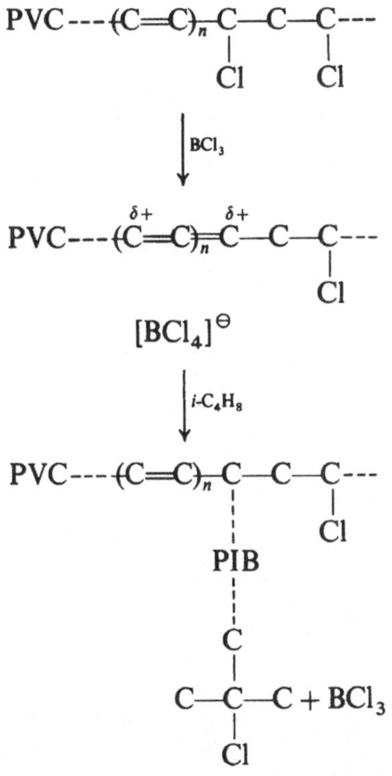

$$PVC\text{---}(C{=}C)_n\text{---}C\text{---}C\text{---}C\text{---}$$
$$\underset{Cl}{|} \qquad \underset{Cl}{|}$$

$$\downarrow BCl_3$$

$$PVC\text{---}(C{=}C)_n\overset{\delta+}{=}\overset{\delta+}{C}\text{---}C\text{---}C\text{---}$$
$$\underset{Cl}{|}$$

$$[BCl_4]^{\ominus}$$

$$\downarrow i\text{-}C_4H_8$$

$$PVC\text{---}(C{=}C)_n\text{---}C\text{---}C\text{---}C\text{---}$$
$$\vdots \qquad \underset{Cl}{|}$$
$$PIB$$
$$\vdots$$
$$C$$
$$|$$
$$C\text{---}C\text{---}C + BCl_3$$
$$\underset{Cl}{|}$$

Table 7 shows some important characteristics of the grafting process and products. According to the data, the yield of isobutylene polymerisation, grafting efficiency (GE), branch frequency (b) and PIB content in the graft increase significantly with increasing concentrations of allylic chlorines in PVC(A).

TABLE 7

EFFECT OF ALLYLIC CHLORINE CONCENTRATION (S, DETERMINED BY OZONOLYSIS) IN PVC(A) ON GRAFTING OF ISOBUTYLENE

Sample	$S = \dfrac{\bar{M}_n^o}{\bar{M}_n} - 1$	$\bar{M}_n \times 10^{-3}$ (PVC)	i-C_4H_8 yield (%)	GE^a (%)	$\bar{M}_n \times 10^{-3}$ (homo-PIB)	$\bar{M}_n \times 10^{-3}$ (PVC-g-PIB)	Branch frequency,^{b} b	PIB in graft (%)
Control	0·11	64·2	6	~2	—	—	—	—
12	0·49	65·2	22	48	49·4	131	1·3	50
13	0·64	64·9	19	61	44·1	138	1·6	53
14	0·88	63·1	29	79	41·2	155	2·2	59
15	1·55	64·9	34	83	38·9	170	2·7	62
16	1·69	63·7	40	89	38·2	200	3·5	68
17	3·70	64·4	64	91	24·1	300	9·7	76

^{a} GE = grafting efficiency.

$$^b\ b = \frac{\bar{M}_n(\text{PVC-g-PIB}) - \bar{M}_n(\text{PVC})}{\bar{M}_n(\text{PIB})}.$$

Subsequent reaction of PVC(A)-*g*-PIB-Cls with Me$_2$CpAl resulted in graft copolymers possessing reactive Cp branch termini, which yielded reversible networks by Diels–Alder/*retro*-Diels–Alder reactions.[22,24]

Grafting of isobutylene on PVC(T)s has also been studied by using Et$_2$AlCl coinitiator. This process yielded poly(vinyl chloride-*co*-2-chloropropene-*g*-isobutylene) (PVC(T)-*g*-PIB) with relatively high grafting efficiencies (~60–70%) and relatively high PIB contents (~60–90%), but low branch frequencies (Table 8).

Since one of the major problems is the stability of PVC and PVC products, thermal and thermooxidative degradation of cyclopentadienylated and grafted PVC(A)s and PVC(T)s have also been studied. Figures 19–24 show the extent of HCl loss (ξ_{HCl}) as a function of time for selected PVC(A)-Cp, PVC(T)-Cp, PVC(A)-*g*-PIB-Cl, PVC(A)-*g*-PIB-Cp and PVC(T)-*g*-PIB samples during thermal and thermooxidative degradation.

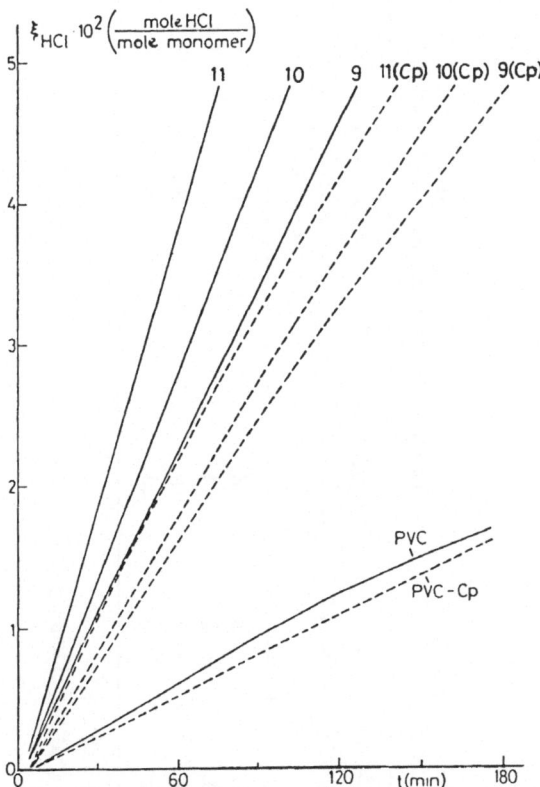

FIG. 19. Extent of HCl loss as a function of time for PVC(A)s and PVC(A)-Cp's during thermal dehydrochlorination (190 °C, N$_2$; for identification see Table 6).

TABLE 8
CHARACTERISTICS OF ISOBUTYLENE GRAFTING ON PVC(T)s

PVC(T) sample[a]	2-CP (mol%)[a]	i-C_4H_8 yield (%)	GE[b] (%)	PIB homopolymer		PVC(T)-g-PIB $\bar{M}_n \times 10^{-3}$	PIB in the graft (%)	b
				$\bar{M}_n \times 10^{-3}$	$\bar{M}_w/\bar{M}_n$			
2	0·78	8	58	101	1·9	180	59	1·05
3	1·42	14	61	111	1·8	188	68	1·16
4	3·45	16	68	98	2·2	285	81	2·36
5	6·46	23	71	139	1·9	400	89	2·65

[a] From Table 4.
[b] GE = grafting efficiency.

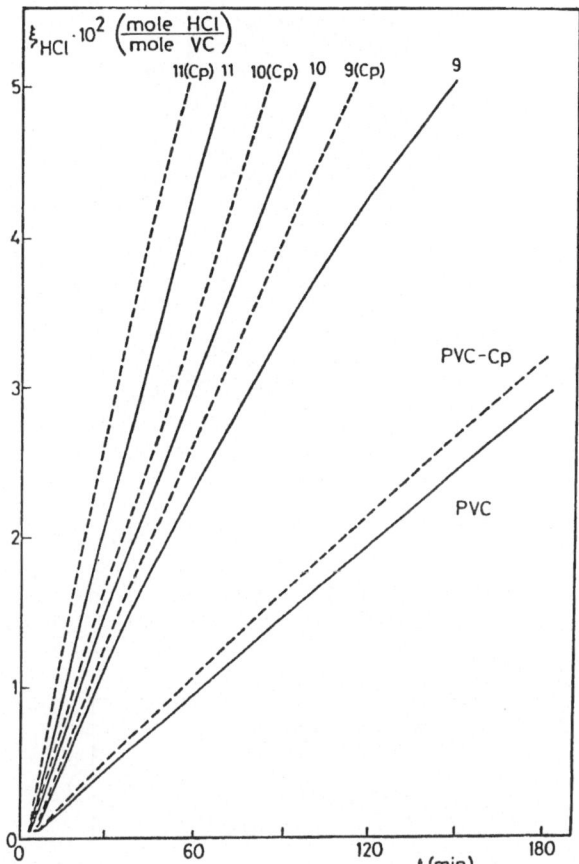

FIG. 20.   Extent of HCl loss as a function of time for PVC(A)s and PVC-Cp's during thermooxidative dehydrochlorination (180 °C, O_2; for identification see Table 6).

According to the data in Fig. 19 the thermal stability of PVC(A)s increases on Me_2CpAl treatment. In contrast, the rates of thermooxidative dehydrochlorination are higher for PVC(A)-Cp than for untreated PVC(A)s, i.e. thermooxidative stability is decreased by the introduction of pendant Cp groups in PVC(A)s (cf. Fig. 20). According to the data in Fig. 21, the rate of HCl loss is lower for cyclopentadienylated PVC(T) (sample 4) during thermal degradation than for virgin PVC(T). However, thermo-oxidative stability markedly decreases on Me_2CpAl treatment of PVC(T).

Figure 22 shows that the thermal stability of PVC(A)s significantly increases upon grafting with isobutylene. Subsequent decreases in rates of

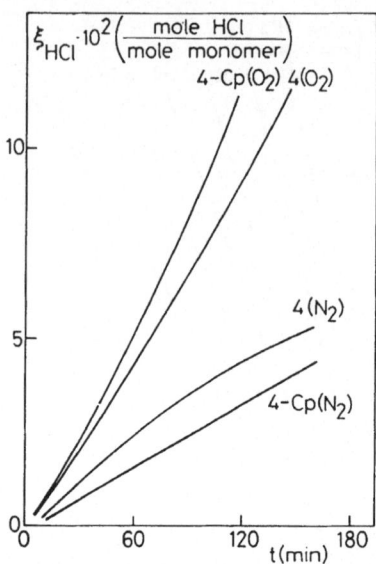

FIG. 21. Extent of HCl loss for PVT(T) (sample 4) and the corresponding PVC(T)-Cp during thermal (N_2) and thermooxidative (O_2) degradation (180 °C).

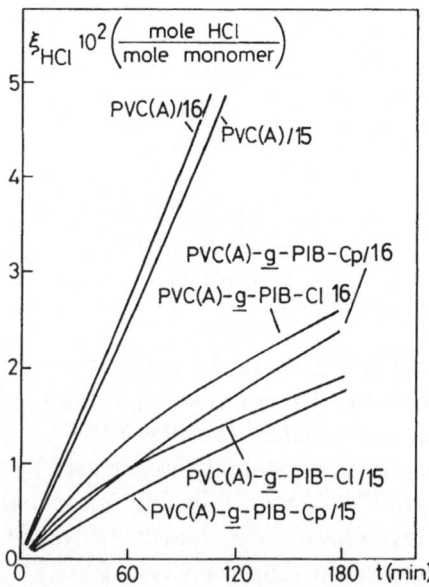

FIG. 22. Extent of HCl loss as a function of time for PVC(A) and the corresponding PVC(A)-g-PIB-Cl and PVC(A)-g-PIB-Cp during thermal degradation (190 °C, N_2).

HCl loss are observed for PVC(A)-*g*-PIB-Cls relative to the corresponding PVC(A)-*g*-PIB-Cps. Figure 23 shows that the thermooxidative stability of PVC(A)-*g*-PIB-Cl is much higher than that of pure PVC(A)s. The thermooxidative stability of PVC(A)-*g*-PIB-Cp is below that of PVC(A)-*g*-PIB-Cl but is still higher than that of PVC(A). According to Fig. 24 both thermal (N_2) and thermooxidative (O_2) stability are markedly increased as a result of grafting of PVC(T) with isobutylene.

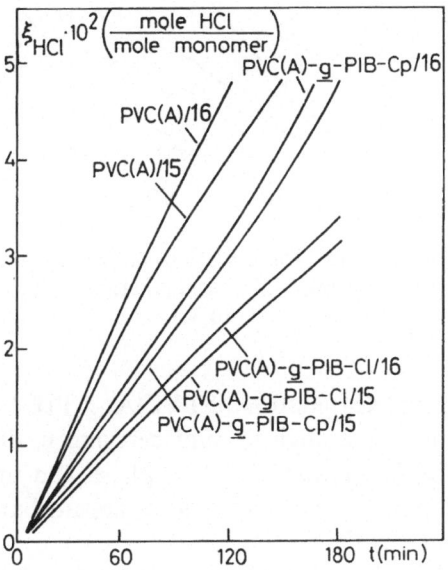

FIG. 23. Extent of HCl loss as a function of time for PVC(A) and the corresponding PVC(A)-*g*-PIB-Cl and PVC(A)-*g*-PIB-Cp during thermooxidative degradation (180 °C, O_2).

The improvement in thermal stability which occurs upon grafting is due to the substitution of labile (tertiary and allylic) chlorines by PIB branches. This finding also demonstrates the low thermal stability of tertiary and allylic chlorines in PVC. According to the plots in Fig. 22, PVC(A)-*g*-PIB-Cl exhibits two dehydrochlorination phases, a relatively fast initial phase followed by a slower subsequent phase. These two phases coalesce on substitution of the tertiary chlorine branch termini by Cp groups (cyclopentadienylation by Me_2CpAl). Tertiary chlorines at PIB branch termini are also unstable[28] and tend to eliminate HCl at a relatively high rate upon heating. Thus, the relatively fast initial dehydrochlorination

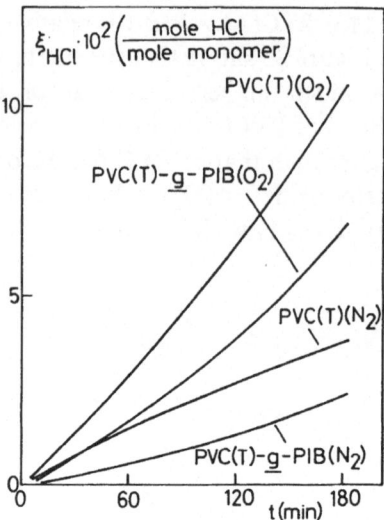

F<small>IG</small>. 24. Extent of HCl loss for PVC(T) (sample 3) and the corresponding PVC(T)-*g*-PIB during thermal (N$_2$) and thermooxidative (O$_2$) degradation (180 °C).

phase during thermal degradation of PVC(A)-*g*-PIB-Cl is most likely due to HCl loss from PIB branch termini carrying tertiary chlorines. Disappearance of the fast initial HCl loss phase and improvement in the thermal stability upon cyclopentadienylation are attributed to a replacement of tertiary chlorines by Cp groups.

The enhanced thermooxidative stability of PVC(A)-*g*-PIB-Cls and PVC(T)-*g*-PIBs relative to the corresponding PVC(A) and PVC(T) samples is also due to a substitution of allylic and tertiary chlorines by PIB branches. This substitution leads to lower rates of initiation of HCl loss and formation of polyenes; reduction in polyene content yields less peroxy radicals by the polyene → peroxide → peroxy radical sequence, and will thus result in a reduced rate of HCl elimination. Following this train of thought, the decrease in the stability of PVC(A)-*g*-PIB-Cl upon cyclopentadienylation may be due to the introduction of highly oxidizable Cp groups in the graft copolymer. Fast oxidation of Cp groups may lead to peroxy radicals which will, in turn, attack regular repeating units in PVC resulting in fast initiation of HCl zip-elimination. These findings also support earlier proposals[4] according to which the main process of thermooxidative degradation of PVC is the polyene → peroxide → peroxy radical sequence in addition to thermal initiation of HCl loss.

6. DEGRADATION OF UNTREATED AND CATIONICALLY MODIFIED LABILE CHLORINE CONTAINING ELASTOMERS

6.1. Thermal Dehydrochlorination of Polyisobutylenes Carrying Tertiary Chlorine Chain Ends

Mono- and difunctional polyisobutylenes (PIB) containing one or two —$CH_2C(CH_3)_2Cl$ termini have been obtained by cationic polymerization. It has been demonstrated that PIBs prepared by the 'H_2O'/BCl_3 initiator system carry one —$CH_2C(CH_3)_2Cl$ endgroup[34] and those synthesized by the p-dicumyl chloride $(Cl(CH_3)_2CC_6H_4C(CH_3)Cl)/BCl_3$ initiator[52,53] system possess two such termini:

$$Cl-\underset{\underset{CH_3}{|}}{\overset{\overset{CH_3}{|}}{C}}-CH_2-\!\!\sim\!\!\sim\!PIB\!\sim\!\sim\!\!-\underset{\underset{CH_3}{|}}{\overset{\overset{CH_3}{|}}{C}}-\!\!\bigcirc\!\!-\underset{\underset{CH_3}{|}}{\overset{\overset{CH_3}{|}}{C}}-\!\!\sim\!\!PIB\!\sim\!\!-CH-\underset{\underset{CH_3}{|}}{\overset{\overset{CH_3}{|}}{C}}-Cl$$

Thermal degradation studies have been carried out with these polymers. Figure 25 shows the extent of HCl loss as a function of time for a difunctional polyisobutylene (Cl-PIB-Cl) in the 170–220 °C temperature range. According to these results thermal dehydrochlorination is complete after ∼30 min at 220 °C and the final value of HCl loss is 2·0 mol

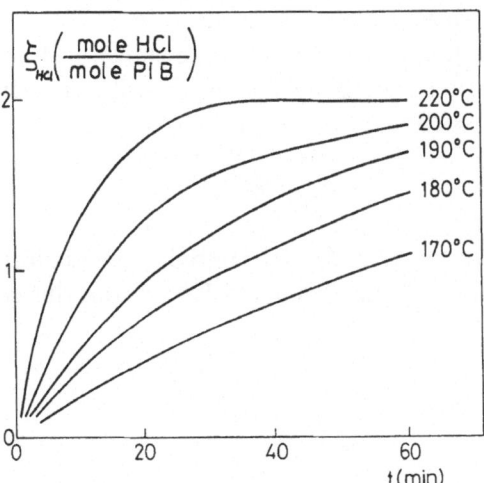

FIG. 25. Extent of HCl loss as a function of time for Cl-PIB-Cl at different temperatures under N_2 atmosphere.

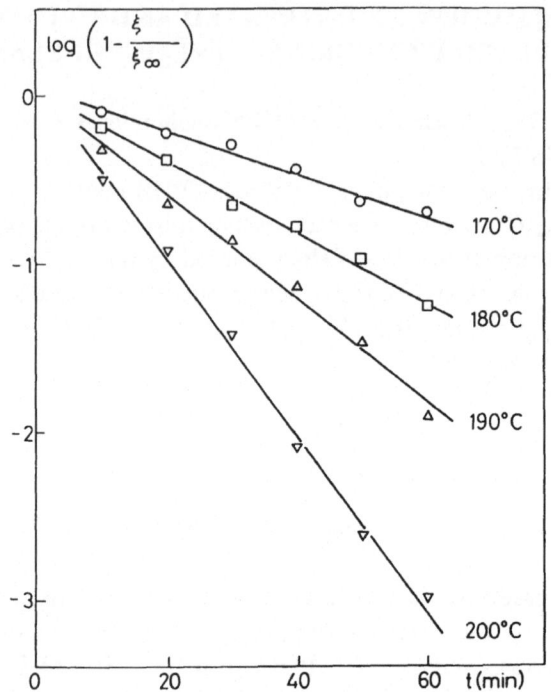

FIG. 26.   Logarithm of $(1 - (\xi/\xi_\infty))$ as a function of time for Cl-PIB-Cl.

$HCl\,mol^{-1}$ PIB. This result proves conclusively that the sample contains two tertiary chlorines per polymer chain.

The corresponding first order plots are shown in Fig. 26. The rate of dehydrochlorination should be first order with respect to HCl in the absence of side reactions:

$$\frac{d\xi}{dt} = k(\xi_\infty - \xi)$$

where k is the rate constant, ξ_∞ is the initial tertiary chlorine content and ξ is the concentration of the eliminated HCl. From this equation

$$\xi = \xi_\infty(1 - \exp(-kt))$$

or

$$\log\left(1 - \frac{\xi}{\xi_\infty}\right) = -kt$$

ξ is available from the ξ versus time curves, while ξ_∞ values were independently determined for every experiment by increasing the temperature after 60–70 min to 220–230 °C and maintaining it in this range for an additional 2 h. The rate constants were determined from the slopes of the $\log(1 - \xi/\xi_\infty)$ versus time plots. The Arrhenius plot of these constants gave a relatively low activation energy, ΔE_a, of 19·1 kcal mol^{-1}. According to low molecular weight mechanistic studies on halide elimination,[54–57] the low activation energy of HCl loss from the tertiary chlorine containing chain end is most likely due to the considerable steric strain at the terminus. A recent study has shown[58] that the dehydrochlorination of Cl-PIB-Cls is complete after 10 h at 190 °C and the resulting olefin can be *endo* or *exo*. This is in very good agreement with results obtained with low molecular weight compounds which may be considered as models, e.g. 2-chloro-2,2,4-trimethyl pentane[57] which yields 81 % 2,4,4-trimethyl-1-pentene and 19 % 2,4,4-trimethyl-2-pentene.

Hydrogen chloride analysis may provide information with regard to the number-average end functionality ($\bar{F}_n$) if $\bar{M}_n$ is known, or can be used as an alternative molecular weight determination method if $\bar{F}_n$ is known:

$$\bar{F}_n = \frac{\bar{M}_n \cdot N}{G}$$

or

$$\bar{M}_n = \bar{F}_n \cdot \frac{G}{N}$$

where N is the number of moles of HCl evolved after complete dehydrochlorination of a polymer of weight G. According to the data in Table 9 $\bar{M}_n$ values obtained by this end group analysis method are in good

TABLE 9

MOLECULAR WEIGHTS OF PIBs DETERMINED BY GPC, OSMOMETRY, ^{1}H-n.m.r. SPECTROSCOPY AND THERMAL DEHYDROCHLORINATION

Sample	$\bar{M}_n \times 10^{-3}$			
	GPC	Osmometry	n.m.r.	Thermal dehydrochlorination
PIB-Cl	11·96	12·3	—	11·52
PIB-Cl	6·63	6·9	—	6·71
Cl-PIB-Cl	7·54	7·6	—	7·53
Cl-PIB-Cl	4·93	5·1	4·74	4·88
Cl-PIB-Cl	2·31	2·4	2·09	2·17

agreement with those obtained by osmometry, GPC and ^{1}H-n.m.r. spectroscopy. These findings quantitatively prove earlier conclusions[34,52,59] with regard to the mechanism of termination in the BCl_3 coinitiated polymerization of isobutylene.

6.2. Thermal Dehydrochlorination of Untreated and Modified Chlorine Containing Elastomers

The nature of labile (tertiary or allylic) chlorines and their concentration (1–2 %) in polychloroprene rubber (CR), chlorinated butyl rubber (Cl-IIR) and chlorinated ethylene–propylene copolymers (Cl-EPM) determine certain important properties of these elastomers, e.g. stability, cure versatility. Although these polymers carry isolated labile chlorines,[67–70] only a few reports have appeared concerning the thermal degradation behaviour of these materials.[4,29–31,60–62]

Figures 27–29 show the extent of HCl loss as a function of time during the thermal degradation of CR, Cl-IIRs and Cl-EPM and those treated with Me_3Al. The untreated materials exhibit initial rapid HCl loss followed by a slower rate. In contrast the polymers modified by Me_3Al show only the relatively slow dehydrochlorination phase. According to Fig. 30, similar findings have also been obtained for cationically grafted and cyclopentadienylated CRs, i.e. chloroprene elastomers grafted with

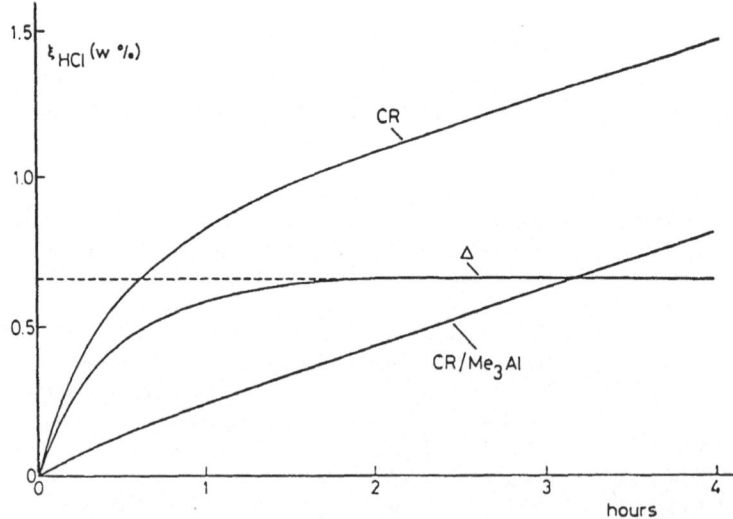

FIG. 27.   Extent of HCl loss as a function of time for CR, CR(Me$_3$Al) and the difference (Δ).

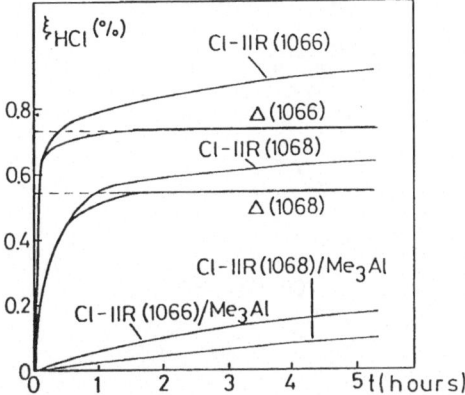

FIG. 28.   Extent of HCl loss as a function of time for Cl-IIR (HT-1066), Cl-IIR (HT-1066) (Me$_3$Al), and Cl-IIR (HT-1068), Cl-IIR (HT-1068) (Me$_3$Al), and the differences (Δ).

isobutyl vinylether (CR-g-PIBVE) and treated with Me$_2$CpAl (CR-Cp). As those cationic techniques (grafting or treatment with Me$_3$Al and Me$_2$CpAl) result in the substitution of allylic or tertiary chlorines, the enhanced thermal stability obtained by these modifications demonstrates the low thermal stability of these chlorines. Similar but much less improvement in stability has been reported for cationically modified PVCs.[21−23,35,36,47,50,63−66] This smaller effect is due to the much lower labile chlorine concentration in PVC than in CR, Cl-IIR or Cl-EPM.

Combination of thermal dehydrochlorination and Me$_3$Al treatment of CR, Cl-IIR and Cl-EPM has led to a new analytical method for the

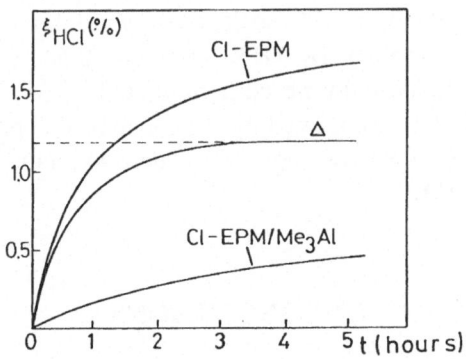

FIG. 29.   Extent of HCl loss as a function of time for Cl-EPM, Cl-EPM (Me$_3$Al) and the difference (Δ).

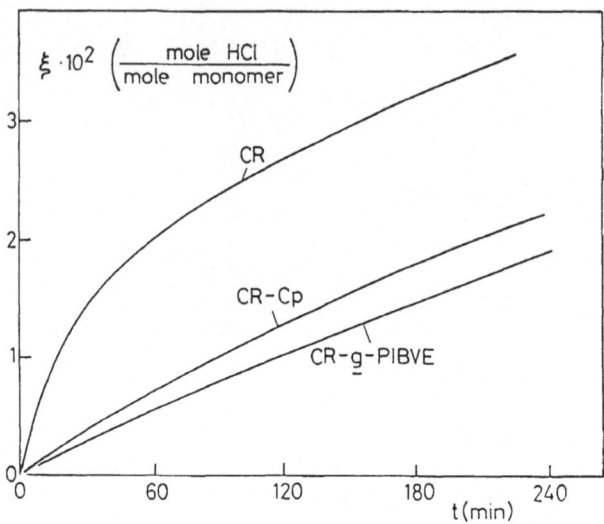

FIG. 30. Extent of HCl loss as a function of time for polychloroprene (CR), cyclopentadienylated polychloroprene (CR-Cp) and poly(chloroprene-*g*-isobutyl vinyl ether) (CR-*g*-PIBVE) during thermal degradation (190 °C, N$_2$).

determination of labile chlorine concentration in these polymers. The difference (Δ in Figs 27–29) between the extent of dehydrochlorination of virgin and Me$_3$Al-treated elastomers at extended degradation times, i.e. during the second stage of HCl loss, yields the labile chlorine content in these polymers:

$$\Delta = \xi_{HCl} \text{ (untreated)} - \xi_{HCl} \text{ (Me}_3\text{Al treated)}$$

According to the data in Figs 27 and 28, the allylic chlorine contents of CR, Cl-IIR (HT-1066) and Cl-IIR (HT-1068) are 0·64, 0·71 and 0·54%, respectively. The labile chlorine content of CR (0·64%) is in excellent agreement with the value obtained by the conventional piperidine method (0·6%). As Fig. 29 shows, our combined method yielded 1·15% tertiary chlorines in Cl-EPM.

7. CONCLUSIONS

On the basis of quantitative results obtained with PVCs containing increased concentrations of labile (tertiary and allylic) chlorines, cationically modified PVCs and elastomers carrying isolated labile chlorines, the primary process of thermal degradation of PVC includes the following steps:

1. initiation at defect sites, randomly at regular monomer units;
2. HCl elimination by zipping (formation of polyene sequences in the chain);
3. termination of zipping.

Systematic studies presented in this review and elsewhere have clearly demonstrated[3,19-24,26,28-31] that allylic and tertiary chlorine containing defects in polymers are thermally unstable and therefore the rate constants for initiation of HCl loss at these sites are relatively high. This conclusion is in accord with the results of other independent studies.[5,11-14,17] The degradation behaviour of low molecular weight model compounds[71-73] also indicates the low stability of allylic and tertiary chlorine containing irregular structures in PVC. It has also been proven quantitatively that initiation of HCl loss takes place not only at defect sites but also randomly at the normal repeat units of the resin.[26] This means that PVC may degrade at significant rates by random initiation at conventional processing temperatures, i.e. the synthesis of a thermally stable PVC seems to be impossible. The rate constant of initiation of HCl elimination at normal repeat units in PVC was found to be four orders of magnitude lower than that at internal allylic chlorines in thermal degradation experiments carried out using an inert solvent at 200 °C. This finding and conclusion do not agree with the results of some model studies,[71] according to which defect-free PVCs are thought to be thermally stable. Evidently, great care has to be exercised when results obtained with low molecular weight model systems are used in order to understand the behaviour of macromolecular systems.

The results discussed in this study and those obtained recently by Haynie et al.[74] contradict the conclusions by Minsker et al.[10,75,76] and Kolinsky et al.[77,78] who claim that tertiary and/or allylic chlorine containing defect sites have no significant effect on the thermal degradation of PVC and that keto-allylic structures (—CHCl—CH=CH—CO—) alone initiate dehydrochlorination and determine the degradation behaviour of the resin. According to Starnes' proposition,[2] however, keto-allyl groups are, most likely, more stable than other defects in PVC because of the electron withdrawing effect of the conjugated carbonyl function.

Anomalous structures and the degradation behaviour of PVC samples obtained under subsaturation conditions (uPVC) have been studied extensively by Braun and Holzer[16,79,80] and Hjertberg and Sörvik.[8,18] The latter have concluded that the thermal stability of uPVCs and of commercial suspension PVCs are associated with tertiary chlorine (Cl_T) containing branch structures,[8,18] and that the activity of allylic chlorine (Cl_A) containing sites is lower than that of tertiary chlorine sites. However, multiple regression analysis of their data is unreliable because of the strong intercorrelation between the concentrations of Cl_A and Cl_T in uPVCs and because the molecular weights (i.e. the concentration of chain end defects) of the samples used are different. Since the absolute values of rate constants for HCl elimination at various defect sites are unknown their data only show a correlation between the rates of HCl loss and anomalous structures in PVC. An accurate method for the determination of the structures and concentrations of chain end irregularities is not yet available.

Unfortunately, only a few acceptable mechanistic studies exist[1,2,6] concerning not only the initiation but also the mechanism of zip-elimination of HCl (polyene formation) and the termination of the process. Our results with PVC(A)s indicate that the rate constant for zip-elimination, i.e. the 'allyl activated' chain propagation of HCl loss, must be orders of magnitude higher than that of initiation at internal allylic chlorines.[26] Thus zip-elimination takes place not by a step-by-step allyl-activated process at the allylic chlorine containing end of a polyene sequence but by an extraordinarily fast zipping process, the mechanism of which still remains obscure. A better understanding of the mechanism of initiation, zip-elimination of HCl and zip termination during thermal degradation of PVC should also lead to a deeper insight into the mechanism of stabilisation of PVC.

Thermooxidative degradation experiments with PVCs containing increased amounts of labile tertiary and allylic chlorines have shown that labile sites strongly influence not only the thermal but also the thermooxidative stability of the resin.[21-24,26] Results obtained with these polymers also support the validity of the 'minimal scheme' of thermo-oxidative degradation of PVC.[4] According to our results, thermal and oxidative initiation of HCl loss are closely coupled to the polyene → peroxide → peroxy radical reaction sequence where the last species would attack normal monomer units of PVC and thus lead to initiation of HCl elimination Additional mechanistic studies are needed to gain a better understanding of the detailed mechanism of thermooxidative degradation and stabilisation of PVC.

REFERENCES

1. DAVID, C., *Comprehensive Chemical Kinetics*, **14** (1975), 1.
2. STARNES, W. H., JR., *Dev. Polym. Degr.*, **3** (1981), 135.
3. IVÁN, B., KENNEDY, J. P., KELEN, T. and TÜDŐS, F., *Polymer Bulletin*, **2** (1980), 461.
4. TÜDŐS, F., KELEN, T. and NAGY, T. T., *Dev. Polym. Degr.*, **2** (1979), 187.
5. BRAUN, D., *Dev. Polym. Degr.*, **3** (1981), 101.
6. TÜDŐS, F. and KELEN, T., *Macromol. Chem.*, **8** (1973), 393.
7. STARNES, W. H., JR., SCHILLING, F. C., PLITZ, I. M., CAIS, R. E., FREED, D. I., HARTLESS, R. L. and BOVEY, F. A., *Macromolecules*, **16** (1983), 790.
8. HJERTBERG, T. and SÖRVIK, E., *Polymer*, **24** (1983), 673.
9. KELEN, T., *Polymer Degradation*, New York, Van Nostrand Reinhold Co., 1983.
10. MINSKER, K. S., KOLESOV, S. V. and ZUYKOV, G. E., *Aging and Stabilisation of Vinyl Chloride Based Polymers*, Moscow, Izd. Nauka, 1982 (in Russian).
11. BRAUN, D. and WEISS, F., *Angew. Makromol. Chem.*, **13** (1970), 67.
12. GUPTA, V. P. and ST. PIERRE, L. E., *J. Polym. Sci.*, *A1*, **8** (1970), 37.
13. BERENS, A. R., *Polymer Preprints*, **14** (1973), 671.
14. BRAUN, D., MICHEL, A. and SONDERHOF, D., *Eur. Polym. J.*, **17** (1981), 49.
15. BRAUN, D. and SONDERHOF, D., *Eur. Polym. J.*, **18** (1982), 141.
16. BRAUN, D. and HOLZER, G., *Angew. Makromol. Chem.*, **117** (1983), 15.
17. AIRINEI, A., BURUIANA, E. C., ROBILA, G., VASILE, C. and CARACULACU, A., *Polymer Bulletin*, **7** (1982), 465.
18. HJERTBERG, T. and SÖRVIK, E., *Polymer*, **24** (1983), 685.
19. IVÁN, B., KENNEDY, J. P., KELEN, T. and TÜDŐS, F., Preprints, *Third Int. Symp. on PVC*, Cleveland, 1980, p. 313.
20. IVÁN, B., KENNEDY, J. P., KELEN, T. and TÜDŐS, F., *J. Polym. Sci., Polym. Chem. Ed.*, **19** (1981), 679.
21. IVÁN, B., KENNEDY, J. P., KELEN, T. and TÜDŐS, F., *Polymer Bulletin*, **6** (1981), 147.
22. IVÁN, B., KENNEDY, J. P., KELEN, T. and TÜDŐS, F., *Polymer Bulletin*, **6** (1981), 155.
23. IVÁN, B., KENNEDY, J. P., KENDE, I., KELEN, T. and TÜDŐS, F., *J. Macromol. Sci.-Chem.*, **A16** (1981), 1473.
24. IVÁN, B., KENNEDY, J. P., KELEN, T. and TÜDŐS, F., *J. Macromol. Sci.-Chem.*, **A17** (1982), 1033.
25. IVÁN, B., TÜDŐS, F., EGYED, O. and KELEN, T., *Makromol. Chem., Rapid Commun.*, **3** (1982), 727.
26. IVÁN, B., KENNEDY, J. P., KELEN, T., TÜDŐS, F., NAGY, T. T. and TURCSÁNYI, B., *J. Polym. Sci., Polym. Chem. Ed.*, **21** (1983), 2177.
27. GUYOT, A., BERT, M., BURILLE, P., LLAURO, M.-F. and MICHEL, A., *Pure Appl. Chem.*, **53** (1981), 401.
28. IVÁN, B., KENNEDY, J. P., KELEN, T. and TÜDŐS, F., *J. Macromol. Sci.-Chem.*, **A16** (1981), 533.
29. IVÁN, B., KENNEDY, J. P., KELEN, T. and TÜDŐS, F., *Polymer Bulletin*, **3** (1980), 45.

30. IVÁN, B., KENNEDY, J. P., PLAMTHOTTAM, S. S., KELEN, T. and TÜDŐS, F., *J. Polym. Sci.*, *Polym. Chem. Ed.*, **18** (1980), 1685.
31. KENNEDY, J. P., PLAMTHOTTAM, S. S. and IVÁN, B., *J. Macromol. Sci.-Chem.*, **A17** (1982), 637.
32. KENNEDY, J. P., *J. Polym. Sci.*, *A1*, **6** (1968), 3139.
33. KENNEDY, J. P., *J. Macromol. Sci.-Chem.*, **A3** (1969), 861.
34. KENNEDY, J. P., HUANG, S. Y. and FEINBERG, S. C., *J. Polym. Sci.*, *Polym. Chem. Ed.*, **15** (1977), 2869.
35. KENNEDY, J. P. (Ed.), General volume on 'Cationic graft copolymerization', *J. Appl. Polym. Sci.*, *Appl. Polym. Symp.*, **30** (1977).
36. GUPTA, S. N., KENNEDY, J. P., NAGY, T. T., KELEN, T. and TÜDŐS, F., *J. Macromol. Sci.-Chem.*, **A12** (1978), 1407.
37. KENNEDY, J. P. and CASTNER, K. F., *J. Polym. Sci.*, *Polym. Chem. Ed.*, **17** (1979), 2055.
38. MICHEL, A., SCHMIDT, G., CASTADENA, E. and GUYOT, A., *Angew. Makromol. Chem.*, **47** (1975), 61.
39. KELEN, T., BÁLINT, G., GALAMBOS, GY. and TÜDŐS, F., *Eur. Polym. J.*, **5** (1969), 597.
40. WIRSÉN, A. and FLODIN, P., *J. Appl. Polym. Sci.*, **22** (1978), 3039.
41. BENGOUGH, W. I. and ONOZUKA, M., *Polymer*, **6** (1965), 625.
42. SCHWENK, U., *Angew. Makromol. Chem.*, **47** (1975), 43.
43. ROBILA, G., BURUIANA, E. C. and CARACULACU, A., *Eur. Polym. J.*, **13** (1977), 21.
44. PERICHAUD, A., DUSSOUBS, D. and SAVIDAN, L., C. R., *Hebd. Seances Acad. Sci.*, *Ser. C.*, **290** (1980), 65.
45. MICHEL, A., BURILLE, P. and GUYOT, A., Preprints, *IUPAC 26th Int. Symp. on Macromol.*, Vol. I, Mainz, (1979), p. 603.
46. STARNES, W. H., JR., PLITZ, I. M., HISCHE, D. C., FREED, D. I., SCHILLING, F. C. and SCHILLING, M. L., *Macromolecules*, **11** (1978), 373.
47. IVÁN, B., KENNEDY, J. P., KELEN, T. and TÜDŐS, F., *J. Polym. Sci.*, *Polym. Chem. Ed.*, **19** (1981), 9.
48. NAGY, T. T., KELEN, T., TURCSÁNYI, B. and TÜDŐS, F., *Polymer Bulletin*, **2** (1980), 77.
49. TÜDŐS, F., KELEN, T., NAGY, T. T. and TURCSÁNYI, B., *Pure Appl. Chem.*, **38** (1974), 201.
50. KENNEDY, J. P. and DAVIDSON, D. L., *J. Appl. Polym. Sci.*, *Appl. Polym. Symp.*, **30** (1977), 13.
51. GUPTA, S. N. and KENNEDY, J. P., *Polymer Bulletin*, **1** (1979), 253.
52. KENNEDY, J. P. and SMITH, R. A., *J. Polym. Sci.*, *Polym. Chem. Ed.*, **18** (1980), 1523.
53. KENNEDY, J. P., CHANG, V. S. C., SMITH, R. A. and IVÁN, B., *Polymer Bulletin*, **1** (1979), 575.
54. BROWN, H. C. and FLETCHER, R. S., *J. Am. Chem. Soc.*, **72** (1950), 1223.
55. BROWN, H. C. and BERNEIS, H. L., *J. Am. Chem. Soc.*, **75** (1953), 10.
56. BROWN, H. C., *J. Chem. Soc.*, 1956, 1248.
57. BROWN, H. C. and MARITANI, I., *J. Am. Chem. Soc.*, **77** (1955), 3607.
58. TESSIER, M. and MARECHAL, E., *Eur. Polym. J.*, **20** (1984), 269.
59. CHANG, V. S. C., KENNEDY, J. P. and IVÁN, B., *Polymer Bulletin*, **3** (1980), 339.

60. JUHÁSZ, K., VARGA, J., DOSZLOP, S. and ZUBONYAI, L., *Plaste Katusch.*, **15** (1968), 636.
61. GARDNER, D. L. and MCNEILL, I. C., *Eur. Polym. J.*, **7** (1971), 569.
62. JOHNSON, P. R., *Rubber Chem. Technol.*, **49** (1976), 650.
63. GAYLORD, N. G. and TAKAHASHI, A., *J. Polym. Sci.*, **B8** (1970), 361.
64. THAME, N. G., LUNDBERG, R. D. and KENNEDY, J. P., *J. Polym. Sci.*, *A1*, **10** (1972), 2507.
65. KENNEDY, J. P. and ICHIKAWA, M., *Polym. Eng. Sci.*, **14** (1974), 322.
66. ABBAS, K. B. and THAME, N. G., *J. Polym. Sci.*, *Polym. Chem. Ed.*, **13** (1975), 59.
67. COLEMAN, M. M. and BRAME, E. G., JR., *Rubber Chem. Techn.*, **51** (1978), 668.
68. YABLINSKII, O. P., SHARMARLIN, V. S., SHITOV, G. A. and CHEKALOVA, E. G., *Khim. Vysokomol. Soedin. Neftekhim.* (1973), 127.
69. BALDWIN, F. P., GARDNER, I. J., MALATESTA, A. and RAE, J. E., Abstracts, *108th Meeting Rubber Division ACS*, New Orleans, 7–10 October 1975.
70. BRUZZONE, M. and CRESPI, G., *Chim. Ind.* (*Milan*), **42** (1960), 1226.
71. ONOZUKA, M. and ASAHINA, M., *J. Macromol. Sci. Rev. Macromol. Chem.*, **C3** (1969), 235.
72. MAYER, Z., OBEREIGNER, B. and LIM, D., *J. Polym. Sci.*, **C33** (1971), 289
73. HOANG, T. V., MICHEL, A., PICHOT, C. and GUYOT, A., *Eur. Polym. J.*, **11** (1975), 469.
74. HAYNIE, S. L., VILLACORTA, G. M., PLITZ, I. M. and STARNES, W. H., JR., *Polymer Preprints*, **24** (1983), 3.
75. MINSKER, K. S., BERLIN, A. A., LISITSKII, V. V. and KOLESOV, S. V., *Vysokomol. Soedin.*, **A19** (1977), 32.
76. MINSKER, K. S., LISITSKY, V. V., KOLESOV, S. V. and ZAIKOV, G. E., *J. Macromol. Sci.*, *Rev. Macromol. Chem.*, **20** (1981), 243.
77. SVETLY, J., LUKAS, R. and KOLINSKY, M., *Makromol. Chem.*, **180** (1979), 1363.
78. SVETLY, J., LUKAS, R., MICHALKOVA, J. and KOLINSKY, M., *Makromol. Chem. Rapid Commun.*, **1** (1980), 247.
79. BRAUN, D. and HOLZER, G., *Angew. Makromol. Chem.*, **113** (1983), 91.
80. BRAUN, D. and HOLZER, G., *Angew. Makromol. Chem.*, **116** (1983), 51.

Chapter 6

KINETICS AND MECHANISMS OF PHOTO-OXIDATION PROCESSES

J. R. MacCallum

Department of Chemistry, University of St Andrews, Scotland, UK

SUMMARY

The process of photo-oxidation of polymers can be rationalised in terms of a chain reaction comprising initiation, propagation, branching, and termination. Aspects of each step are discussed and the particular problems associated with solid-state reactions are highlighted. The distribution of reactants and products, the significance of diffusion controlled steps, and the role of trace impurities are considered. The relevance of such processes to stabilisation and prediction of material life time is briefly investigated.

1. INTRODUCTION

The area of polymer science covered by what is an apparently simple title is extensive and complex. The result of photo-oxidation of plastics has an economic impact on the application of materials in a variety of environments, while the study of the kinetics and mechanism provides a fascinating field for investigation of solid-state reactions. The ultimate goal, in which the background research combines with applied requirements, is the stabilisation of plastics coupled with the facility for prediction of useful life time.

The purpose of this article is not to make a detailed review of the subject; that objective has already been fulfilled by a number of excellent contributions which cover the extensive literature on photo-oxidation.[1-7] Rather it is intended that specific facets of the complex process should be high-lighted, and perhaps given more prominence than hitherto. In particular, implications of the effects of high viscosity on kinetics and mechanism will be emphasised.

The kinetic processes involved in photo-induced oxidation result in a branching chain reaction which comprises the four normal stages: initiation, propagation, branching and termination. Obviously a knowledge of the mechanism and rates of each of the composite steps would lead to a full understanding of the overall reaction. However, as with almost all branching chain reactions the situation is complex and development of a mathematical expression capable of direct integration and consequently applicable for predicting long-term behaviour of plastics is slow.

It is convenient to attempt to construct a view of the total mechanism before attempting to evolve mathematical relationships. This approach is appropriate since the relevance of the final conclusions must be related to the validity of the proposed mechanism.

2. MECHANISM

2.1. Initiation
A given polymer may be subjected to a variety of treatments before finally being exposed to radiation. There are few samples which can be accurately described as pure. A number of sources of impurities can be defined and any such intrusion which absorbs incident radiation may be an initiator.

2.1.1. Catalyst Fragments
Generally free-radical initiators of polymerisation produce chemically incorporated fragments which are relatively stable on irradiation. Isolated double bond units absorb high energy radiation not found in the solar output at the earth's surface ($\lambda > 300$ nm). However, it is worth noting that oxygen can be copolymerised by a free-radical mechanism to produce a peroxy main-chain link.[8] For example, polystyrene synthesised by free-radical initiators in the presence of O_2 yields the following:

$$\sim CH_2-\underset{\underset{C_6H_5}{|}}{CH}-O-O-CH_2-\underset{\underset{C_6H_5}{|}}{CH}\sim$$

Peroxides and hydroperoxides absorb in the region 300–360 nm.

Materials synthesised by the use of transition metal catalysts almost always contain significant amounts of transition metal ions which may initiate photo-oxidation.

2.1.2. Impurities Arising from Pre-Treatment

Test samples of a polymer are prepared either by casting from solution or hot-pressing the molten material. Both techniques introduce chromophores capable of photo-initiation of oxidation. In the former case residual solvent may undergo reaction[9] and in the latter metal ions introduced by way of surface contact can catalyse the formation of hydroperoxide groups. A further possible route is rupture of carbon–carbon main-chain bonds caused by strain introduced by cooling or crystallisation.

In summary it can be stated that the species responsible for primary photo-initiation is hydroperoxidic in nature and the reaction can be written

$$ROOH + hv \longrightarrow RO^{\cdot} + \dot{O}H \tag{1}$$

The extinction coefficient of hydroperoxides is of the order of 10 and consequently the rate of initiation is very slow.

The occurrence of a not infrequently proposed step:

$$RH + hv \longrightarrow R^{\cdot} + {}^{\cdot}H$$

is highly improbable.

2.2. Propagation

The alkoxy and hydroxy radicals produced by photolysis of hydroperoxides are both reactive species and can rapidly abstract hydrogen atoms, thus:

$$\begin{matrix} RO^{\cdot} \\ HO^{\cdot} \end{matrix} + 2PH \longrightarrow 2P^{\cdot} + ROH + H_2O \tag{2}$$

Generally, alkyl radicals are very reactive, adding O_2 to form peroxy radicals:

$$P^{\cdot} + O_2 \longrightarrow PO_2^{\cdot} \tag{3}$$

The resultant radical, although fairly stable depending on the nature of P, will abstract hydrogen to complete the propagation process and regenerate $P^{\cdot}$:

$$\dot{P}O_2 + PH \longrightarrow PO_2H + P^{\cdot} \tag{4}$$

$$O_2$$

The product of this propagation cycle is hydroperoxide which can undergo further reaction by photolysis.

2.3. Chain Branching

The chain branching step is:

$$POOH + h\nu \longrightarrow PO^{\cdot} + {}^{\cdot}OH \qquad (5)$$

and is identical to the primary initiation mechanism. Alkoxy radicals may react in other ways than simply hydrogen abstraction. The reaction paths followed by $PO^{\cdot}$ can best be summarised by:

$$2PO^{\cdot} + O_2 \longrightarrow PC{=}O + PO_2^{\cdot} \qquad (6)$$

A consequence of this reaction is the formation of ketones which are much better absorbers of radiation than hydroperoxides. The photochemistry of ketones is complicated; however, further peroxy radicals can be produced:

$$PC{=}O + h\nu' \longrightarrow aPO_2^{\cdot} \qquad (7)$$

where a lies between 2 and 0.

2.4. Termination

In the absence of added stabiliser it is not possible to propose a feasible linear termination step which therefore proceeds bimolecularly involving only two radical species. Reaction (3) in propagation is also bimolecular and is fast for the type of radicals normally produced in a polymer matrix. For this reason the termination reaction can be written:

$$RO_2^{\cdot} + R\dot{O}_2 \longrightarrow ROOR + O_2 \qquad (8)$$

The products of reactions involving two peroxy radicals have been extensively studied and the mechanism proposed is the most probable, with the oxygen so formed being in the singlet state, a species which reacts readily with enes to yield hydroperoxides.[10]

The above equations outline the basic mechanism of photo-oxidation of polymers. Details of reactions of particularly the alkoxy radicals can be omitted, at least in a preliminary analysis, without seriously affecting the model. The obvious development would be now to carry out a kinetic analysis using the wealth of rate constant data for liquid phase reactions available in the literature. However, such a procedure would be misleading if not totally erroneous. The next stage in the development of a kinetic scheme capable of evaluation and comparison with experimental observations is to examine, at least qualitatively, the effect of viscosity on the proposed scheme.

3. UNIMOLECULAR REACTIONS

Unimolecular decomposition occurs on direct absorption of photons or on rearrangement of, for example, alkoxy radicals. The consequence of caging radical products is not of great significance in the liquid phase. However, the products of scission of a hydroperoxide group are capable of reacting thus:

$$POOH \xrightarrow{hv} (PO^{\cdot} + {}^{\cdot}OH)_{cage} \longrightarrow PC{=}O + H_2O \qquad (9)$$

For peroxides the equivalent caging effect would be:

$$POOP \xrightarrow{hv} PO^{\cdot} + PO^{\cdot} \longrightarrow PC{=}O + POH \qquad (10)$$

The result of such a cage reaction would be to diminish the efficiency of primary initiation and transfer the chromophore predominance to ketone functions. It is also worth noting that this reaction produces water which, as is noted later, plays a part in the overall process and which has been observed as a major component in volatile products.[11]

Rearrangement of alkoxy radicals as shown in reaction (6) is an important step for producing ketones. For example, in the photo-oxidation of polypropylene the following sequence of reactions can be postulated:

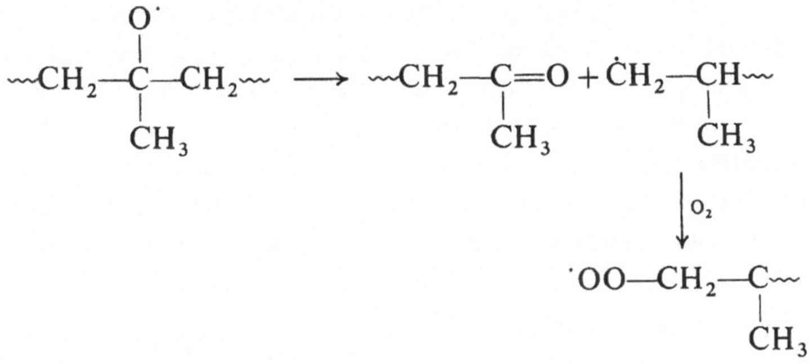

Reaction (11) is in competition with reaction (2) but medium restrictions on the latter may enhance the contribution of the former.

4. BIMOLECULAR REACTIONS

In mobile liquid phase reactions it is assumed that reactants and products are uniformly distributed throughout a given sample. In the event that this basic condition is not met then special problems require to be solved before detailed kinetic analysis can be completed. In this part of the review the significance of bulk viscosity on mechanism will be highlighted and subsequently the kinetic consequences will be dealt with.

The rate of a particular step depends on the concentration of reactants as well as the rate constant. The bimolecular processes which contribute most to the mechanism are reactions (2), (3), (4) and (8). Rate expressions involving [PH], reactions (2) and (4), are less likely to be seriously subject to viscosity control than the expressions for reactions (3) and (8).

A rather interesting concept has developed for mobility of reactive sites—that of relay processes. This can be illustrated as follows:

$$P^{\cdot} + PH \longrightarrow PH + P^{\cdot}$$

Essentially the result of the relay is to translate the radical species through the sample. This transfer reaction is in competition with reaction (3) and in circumstances in which the concentration of O_2 is severely diminished the relay may become operative. In a very stiff medium the free spin would be likely to be localised between the two groups involved since reorientation by bond relaxation would be slow and caging would be effective.

An important outcome of restricted molecular mobility is that products of reaction will themselves be localised around the initial site at which the chain commenced. This fact will have a major effect on the rates of the constituent steps by operating on the concentration dependent part of the rate expression. Allowance for viscosity effects on the mechanism of the chain reaction is not an important matter. However, molecular and segmental mobility will profoundly affect the kinetics. Qualitatively restricted motion will lower a rate by diminishing the rate constant; on the other hand, rates will be increased by higher localised concentrations of products which as a result of chain branching become major contributors to photo-oxidation.

5. RADICAL REACTIVITY

In outdoor applications a plastic is exposed to a mixture of gases during irradiation. The role of O_2 in photo-decomposition of polymers requires no comment; however, the presence of water vapour either as a product of oxidation or arising from the humidity of the atmosphere may well be a relevant parameter in determining radical reactivity. It has been shown that peroxy radicals can complex with hydrogen donating molecules and thereby change their reactivity.[12] Thus, when water is present in a polymer matrix undergoing photo-oxidation the following equilibrium may be established:

$$PO_2^{\cdot} + nH_2O \rightleftharpoons (PO_2^{\cdot} \cdots nH_2O) \qquad (12)$$

The solvated radical has a lower activity than its precursor.

Radicals have also been shown to complex with metal ions and aromatic nuclei both of which are found in, for example, stabilisers. As yet, not enough is known about the relationship between reactivity and complexation to allow reliable prediction of behaviour. Illustration of the effect of transition metal ions is available from work carried out by Tkac.[13]

As far as photo-oxidation of polymers is concerned the principal conclusion of general applicability which derives from the work of Tkac[13] and Halpern[14] is that a transition metal redox cycle effectively allows storage of free-radicals.

6.   LONG-RANGE INTERACTIONS

The significance of bulk viscosity on the kinetic rate constant and spatial distribution of products has been discussed. In a situation in which collisional activation involving reactant species is restricted, other long range interactions may play a role. Electronic excitation energy can be passed from a donor to acceptor over separation distances of up to 8 nm.[15] Details of mechanisms of energy transfer are not appropriate to this review but a good example of the process can be found in photo-oxidation which has progressed beyond the primary stage of building up a population of O–O–H groups. Although they absorb very little of the higher energy of the solar spectrum, hydroperoxides decompose quantitatively on photo-excitation ultimately to produce ketones. The product ketones, having relatively higher extinction coefficients, become the predominant photo absorbing species but do not react quantitatively to yield radicals. It has been shown however that hydrogen bonding of hydroperoxides to ketones can occur and in this case electronic energy transfer can occur very efficiently from excited ketone to hydroperoxide with concomitant breakdown of the latter.[16] In this way ketones act as sensitisers for the decomposition of the poorly absorbing hydroperoxides.

7.   ALTERNATIVE REACTIONS

The scheme outlined above accounts for the main features of photo-oxidation reactions of polymers, that is, development of carbonyl and hydroperoxides accompanied by either main chain scission or crosslinking resulting in a significant change in the physical properties of the material. Experimentally it is common to assess the extent of reaction by measurement of the carbonyl index of a sample. This technique requires a considerable amount of oxidation before it can be applied. An alternative approach is to measure the consumption of gaseous oxygen by a sample undergoing irradiation. This method is very much more sensitive than that involving infrared determination of carbonyl groups and can be used for studying the very early stages of photo-oxidation.

A very sensitive technique which is particularly useful for following changes at or near the surface of a sample is electron scanning for chemical analysis (ESCA). Some very interesting results have been obtained by Clark and co-workers using this apparatus.[17]

Measurements of gas uptake by manometric technique show that significant quantities of CO_2 and H_2O are evolved from a sample undergoing photo-oxidation.[11] A large fraction of the oxygen consumed can be replaced by CO_2 and H_2O. The scheme outlined above does not account for the formation of CO_2 and one possible route is as follows:

8. KINETICS

A relatively simple mechanism has been outlined above (reactions (1), (2), (3), (4) and (8)) and a kinetic expression can be produced for comparison with experimental results. A few reasonable assumptions must be made; for example, the steady state is assumed and also an initial concentration of hydroperoxide species. On this basis Karpukhin and Slobodetskaya derived the following expressions:[18]

$$[POOH] = \frac{k_4^2[PH]^2}{k_8 Q} [1 - \exp(-Qt/2)]^2$$

$$[C{=}O] = \frac{k_4^2[PH]^2}{k_8 Q} [Qt - (3 - \exp(-Qt/2))][1 - \exp(-Qt/2)]$$

in which $[POOH]$ and $[C{=}O]$ are, respectively, the concentrations of hydroperoxide and carbonyl functions at time t, and Q is the quantum yield for decomposition of hydroperoxide to radical species. Qualitatively, the expressions demonstrated the behaviour observed for polypropylene undergoing photo-oxidation but it is clear the mechanism is oversimplified. For example, photolysis of the ketone groups is not included. Unfortunately, enlarging the mechanism makes solution of the differential equations impossible without use of extensive computation. This development is now taking place.[19] The fact that the above workers observed good qualitative agreement with experimental results indicates that a more complex mechanism coupled with sophisticated methods of calculation may result in quantitative agreement. However, a number of aspects of this kinetic analysis require further comment since many of the concepts used are more commonly applied to gas and liquid state kinetics and translation to solid/viscous medium applications is a procedure which cannot automatically be accepted.

8.1. Rate Constants

For a bimolecular reaction having rate constant k_8 it can be shown that when the viscosity of the medium reaches such a value that the diffusive displacement of reactants becomes rate controlling then k_8 can be expressed thus:

$$k_8 = \pi 4 N D R$$

where N is Avogadro's number, D is the composite diffusion coefficient and R is the distance between the reactive centres when reaction occurs. This relationship has been derived assuming a random distribution of reactants, a situation which most probably does not prevail for a polymer in solution, and a constant value for D. A cursory examination of the changes taking place in a polymer undergoing photo-oxidation immediately indicates that changes in the bulk viscosity of the medium brought about by chain scission, or more commonly, crosslinking will be reflected in the bimolecular rate constant through D. For example, in a photo-oxidation resulting in crosslinking, as reaction proceeds D will decrease, k_8 will decrease, and thus, for example, in the above mechanism the rate of termination by bimolecular reaction of peroxy radicals will decrease. A diminution of the rate of termination will increase the chain length of oxidation and thereby increase the rate of crosslinking. This qualitative argument demonstrates how a type of autocatalysis can be established. Little is known about the relationships which govern how diffusion coefficients of large, or even small, molecules vary with extent of crosslinking.

8.2. Concentration of Oxygen

Reaction (3), the addition of oxygen to a radical, is assumed to predominate over all other possible reactions the macroradical could undergo. This step is also bimolecular and what applies in Section 8.1 above is relevant for this reaction. Normally, the diffusion coefficient for O_2 in a polymer is very much higher than that for larger molecules, and the restrictions implied in Section 8.1 are less severe for the small molecules. However, a related problem arises for thick samples which suffer deep penetration of radiation. In this event radicals formed during initiation rapidly consume O_2 by reaction (3). The O_2 thus removed must be replaced by O_2 diffusing upward from deeper parts of the sample or down through the exposed surface. For high rates of production of radicals associated with a combination of high concentrations of chromophore (advanced state of oxidation) and high doses of radiation (accelerated testing procedures), it is possible to convert a photo-oxidative degradation to a mixed photolysis/photo-oxidation reaction. In photolysis the reactivity of R˙ rather than ROO˙ species becomes dominant. Periods of low, or no, radiation allow re-establishment of a uniform concentration of oxygen. In mixed reactions studied by measurement of O_2 uptake it is common to observe quite significant dark reactions as trapped R˙ reacts with O_2 which diffuses into the sample.[20]

8.3. Concentration

As is normal in the analysis of a kinetic scheme, rate expressions are formulated in terms of concentrations of reactants. This usage implies a random distribution of reacting species throughout the sample; failure to achieve this distribution will require modification to the analysis. Normally in a low viscosity medium reactants and products mix to a uniform concentration by rapid diffusion. However, when reactions occur in highly viscous media this randomisation process becomes inefficient and highly localised concentrations develop. In the extreme case it is possible to envisage a situation in which initiation takes place at a particular site and subsequent oxidation occurs around this site. In many ways this model is akin to aspects of solid-state decomposition of crystals and one method of analysis would be to consider the oxidation process in terms of growing, and ultimately overlapping, spheres of reacted polymer. Such an approach has not been published but attempts to develop this model have indicated that interesting results can be simulated in fairly elementary calculations.[21]

8.4. Accumulation of Products

The reasoning behind the ideas proposed in this section are based on the basic fact already discussed in Sections 8.1–8.3—the reaction of photo-induced oxidation takes place in a very viscous medium. The primary product of photo-oxidation is a hydroperoxide, and secondary products are ketonic and peroxidic in nature. All of these products absorb the incident radiation normally found at the earth's surface. Accumulation of products around the site of initiation gives rise to the type of localised reaction discussed in Section 8.3. A number of physical properties of the products can determine their mobility. Transition metals readily form complexes with hydroperoxides[22] which themselves hydrogen-bond to available polar groups such as acids or ketones. In a non-polar hydro-carbon the solubility of polar molecules is limited and once the solubility is exceeded, particularly in localised reactions, the molecules will tend to form insoluble molecular aggregates which consequently will be extruded from the bulk of a sample to its surface.

8.5. Attenuation of Radiation

In laboratory investigations conditions are usually chosen such that only a fraction of incident radiation is absorbed by a sample undergoing examination. In this way it is correctly assumed that the photon flux absorbed is uniform throughout the sample. For the simplified scheme given above the overall rate of reaction will not be directly proportional to the intensity of absorbed radiation. Furthermore, with two branching chain processes resulting from absorption by hydroperoxide and ketone functions the quantum yield of oxidation will vary with the wavelength of incident light. Thus, the desirable condition of low absorption of monochromatic radiation can be arranged in a laboratory experiment, although the immediate consequence is an experiment requiring a long time to perform. For samples undergoing accelerated test, that is, high intensity of polychromatic absorbed radiation, estimation of extent of attenuation of radiation poses a severe problem. In normal conditions of exposure, during weathering trials for example, the depth of penetration and the kinetic consequences of a quadratic termination step could significantly alter the characteristics of a photo-oxidation experiment. The addition of stabilisers giving mixed linear and quadratic termination makes prediction of kinetic performance a daunting task.

8.6. Light and Dark Periods

This section is related to Section 8.5 in that the phenomenon considered is relevant to the comparison of results obtained in test conditions with results obtained under corresponding exposure trials. In the outdoor use of plastics the level of radiation experienced is low and highly polychromatic. Rates of chain initiation by direct photolysis or by chain branching are relatively slow with the result that concentrations of reactive intermediates are very low and chain lengths long. The effects of control of distribution of reactants and products by diffusive processes will be relatively less important than, for example, in higher intensity laboratory experiments. Thus, establishment of uniform concentrations of reactants, including stabilisers, may not be rate controlling and, equally important, establishment of equilibria involving complexation will be complete. The occurrence of a long dark period will facilitate these 'equilibration' processes with consequent effects on mechanisms. For example, accumulation of products around sites of initiation may well play little part in photo-oxidation carried out at low intensity of radiation. Another important reaction which can proceed is the decomposition/stabilisation of hydroperoxides by complexation with added transition metal ions.[22]

9.   STABILISATION OF POLYMERS

A vast amount has been written about the ways in which the chain reaction comprising photo-oxidation can be minimised, if not halted completely. It is apparent that a good understanding of the basic mechanism will be necessary for planned development of a stabiliser system. The means may be obvious but the end difficult to achieve. Removing initiating centres, terminating the chain, preventing chain branching, are all readily identifiable ways for stopping or markedly slowing down the reaction. Successfully performing these functions is not simple, however.

It is not intended that the field of stabilisation should be reviewed and re-assessed: however, it is worth stating that many of the points made earlier in this review must be taken into account when postulating a mechanism of operation for a particular stabiliser. In particular, the importance of medium viscosity for bimolecular reactions should always be borne in mind, along with the concept of random distribution of reactants and products.

Some features of transition metal chemistry are notable. Their role in initiation and their ability to stabilise radicals are features which are yet to be fully investigated. A great deal is known about the photophysics and photochemistry of a wide range of stabilisers but the corresponding properties of stable radicals, particularly peroxy radicals, have received little attention and yet in a polymer matrix the concentration of such species can be considerable in comparison to analogous fluid solutions.

It would be inappropriate to review the field of photo-oxidation of polymers without making reference to the class of stabiliser known as hindered amine light stabilisers (HALS). These compounds function as good stabilisers and yet their mechanism of operation remains a matter of some controversy. The amine itself operates as a stabiliser, as do derived nitroxyl, hydroxylamine and hydroxylamine ethers, which are products of the oxidation of the amine.[6] One possible way in which these compounds function is as chain breaking agents but, as has been clearly demonstrated by Scott,[23] a further regenerative step is required for completion of the cycle. A mechanistic difficulty arises in a sample in which the reactants are randomly distributed, for in that case the occurrence of two bimolecular steps by a single species would be improbable. The Scott mechanism almost certainly applies to a system involving highly localised concentrations of reactants. Indeed the nature of the polymer matrix could be a controlling factor in determining how the reactants are distributed and consequently the mode of operation of HALS may change in different materials. The important point to note is that the mechanism of action of a stabiliser could be rate *and* matrix dependent.

10.   CONCLUSION

The mechanism of photo-oxidation of polymers is a highly complex process both in a chemical and a physical sense. The ultimate objective of predicting the performance of a particular material under specified conditions is still a long way from achievement. A start to this goal has been made using relatively simple models and making simplifying assumptions. The development of sophisticated computing procedures has allowed real progress to be made. It is worth noting that computation of extent of photo-oxidative decomposition of a polymer has shown agreement with experimental observations using a mechanism comprising over 35 individual steps and numerically integrating the relevant differential rate expressions.[19]

REFERENCES

1. CARLSSON, D. J. and WILES, D. M., *J. Macromol. Sci., Rev. Macromol. Chem.*, **14** (1976), 65.
2. ARNAUD, R. and LEMAIRE, J., In: *Developments in Polymer Photochemistry—2*, ed. N. S. Allen, London, Applied Science Publishers, 1981, p. 135.
3. ALLEN, N. S., In: *Developments in Polymer Photochemistry—2*, ed. N. S. Allen, London, Applied Science Publishers, 1981, p. 235.
4. POSPISIL, J., In: *Developments in Polymer Photochemistry—2*, ed. N. S. Allen, London, Applied Science Publishers, 1981, p. 53.
5. BILLINGHAM, N. C. and CALVERT, P. D., In: *Developments in Polymer Stabilisation—3*, ed. G. Scott, London, Applied Science Publishers, 1980, p. 139.
6. DOGONNEAU, M., IVANOV, V. B., ROZANTSEV, E. G., SHOLLE, V. D. and KAGAN, E. S., *J. Macromol. Sci., Rev. Macromol. Chem.*, **22** (1983), 169.
7. SEDLAR, J., MARCHAL, J. and PETRIY, J., *Polym. Photochem.*, **2** (1982), 175.
8. BOVEY, F. A. and KOLTHOFF, I. M., *J. Amer. Chem. Soc.*, **69** (1947), 2143.
9. EASTON, M. J. and MacCALLUM, J. R., *Polym. Degrad. and Stabil.*, **3** (1981), 229.
10. RÅNBY, B. and RABEK, J. (Eds), *Singlet Oxygen*, New York, John Wiley, 1978.
11. BOUSQUET, J. A. and FOUASSIER, J. P., *Polym. Degrad and Stabil.*, **5** (1983), 113.
12. ANDRONOV, L. M., MAIZUS, Z. K. and ZAIKOV, G. E., *Zhur. Fiz. Khim.*, **41** (1967), 1122.
13. TKAC, A., *Int. J. Radiat. Phys. Chem.*, **7** (1975), 457.
14. HALPERN, J., *Acc. Chem. Res.*, **3** (1970), 386.
15. FÖRSTER, T., *Disc. Faraday Soc.*, **27** (1959), 1.
16. GEUSKENS, G. and DAVID, C., *Pure and Appl. Chem.*, **51** (1979), 2385.
17. CLARK, D. T., DILKS, A. and THOMAS, H. R., *Developments in Polymer Degradation—1*, ed. N. Grassie, London, Applied Science Publishers, 1977, p. 87.
18. KARPUKHIN, O. N. and SLOBODETSKAYA, E. M., *J. Polym. Sci.*, A-1, **17** (1979), 3687.
19. GUILLET, J. E., University of Toronto, unpublished work.

20. MacCallum, J. R. and Ramsay, D. A., *Europ. Polym. J.*, **13** (1977), 945.
21. MacCallum, J. R., University of St Andrews, unpublished work.
22. Black, J. F., *J. Amer. Chem. Soc.*, **100** (1978), 527.
23. Bagheri, R., Chakraborty, K. B. and Scott, G., *Polym. Degrad. and Stabil.*, **5** (1983), 145.

Chapter 7

KINETICS AND MECHANISM OF THE OXIDATION OF STRESSED POLYMER

N. Ya. Rapoport and G. E. Zaikov

Institute of Chemical Physics, Academy of Sciences of the USSR, Moscow, USSR

SUMMARY

The fundamental property of chemical reactions in solid polymers is the dependence of reaction kinetics on the structural and molecular-dynamical parameters of a matrix. The latter may differ significantly under stress. Current data concerning the kinetics of oxidation of stressed polymers are reviewed in terms of the relation between stressed polymer structure and molecular motion on the one hand and chain radical reaction kinetics on the other. The influence of stretching stress on different steps in the chain autoxidation process and the relation between oxidation kinetics and polymer durability are discussed for the oxidation of polyolefines.

1. INTRODUCTION

In service, polymer materials have as a rule to withstand stresses of external and internal origin. The latter may arise either during the processing of the polymer or during ageing. Thus, the chemical reactions leading to ageing of polymers usually take place in stressed polymer matrices.

It is a fundamental property of chemical reactions in solid polymers that the chemical kinetics depend on the structural and physical parameters of the matrices which may differ appreciably depending on whether the matrix

207

is stressed or not. This implies that the chemical kinetics of processes occurring in unstressed and stressed polymers will be different. Neither the lifetime of a polymeric material can be predicted nor can an optimum method of stabilising it be determined, unless the effect of stresses on the kinetics and mechanism of ageing has been assessed and taken into account.

The first paper to demonstrate the promoting effect of stresses on the rate of oxidation of vulcanised rubber (which the authors described as 'mechanical activation'[1,2]) appeared over 30 years ago. Since then, a wealth of diverse information, illustrating the effect of stresses on the kinetics of oxidation of polymers of various kinds with molecular oxygen and ozone, has been acquired. This review is an attempt to analyse and generalise the available data.

Scheme 1 illustrates some of the reasons for the relationships between the kinetics of reactions in solid polymers and their structural and physical parameters.

The structural microscopic heterogeneity of solid polymers has important kinetic implications,[3-5] even if the polymer is amorphous. First, the micro-nonuniformities in reactant distribution through the matrix should

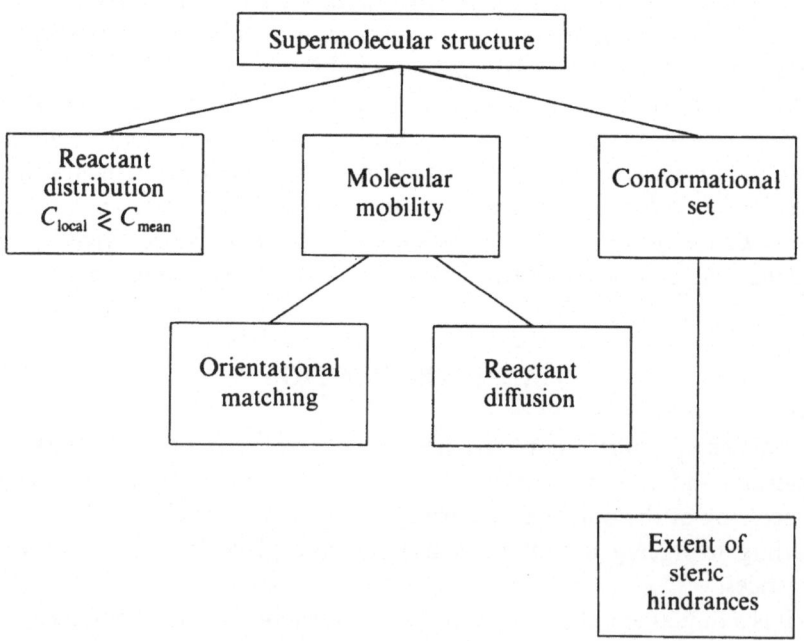

SCHEME 1.

be mentioned. Because of these, the local reagent concentrations may differ greatly from the mean values,[6,7] and the rate constants determined experimentally may not agree with the true values. The micro-nonuniformity of reactant distribution may also give rise to polychromatic kinetics which is typical of reactions in solid matrices: reaction will be faster in the zones of higher local concentrations and as such zones are used up, zones of the next highest concentration will become conspicuous. Formally, this may be described as a distribution according to rate constants.

The micro-nonuniformity of reactant distribution exists in its most explicit form in semicrystalline polymers. For example, the stable nitroxyl radical is contained in polypropylene (PP) only in the amorphous phase.[6] Antioxidants (AO) are also concentrated in the amorphous regions. Their solubility in isotropic PP is determined by the supermolecular structure (SMS) of the polymer.[7] Thus, about 80% of the AO is dissolved in spherulites, in interfibrillar and defective intrafibrillar regions, whereas 20% is confined to spherulite interfaces.[7] The nonuniform distribution of AO through the polymer matrix should always be given due consideration in the practice of polymer stabilisation.

The second important structural and physical parameter influencing the kinetics is the distribution of frequencies and amplitudes of molecular motions. This leads to a distribution of rate constants which has been discussed in detail elsewhere.[8] The relationship between kinetics and molecular mobility is due, first, to the way in which the molecular dynamics of the polymer affect the rate of encounter of reactants during diffusion and, second, to the direct effect of molecular dynamics on an elementary reaction (cf. refs 6, 9, 10). The possible physical mechanisms of the effect of molecular motion on reactivity are discussed elsewhere.[8]

Finally, specific to reactions in polymers is the relationship between chemical kinetics and the conformation of the polymer which stems from the fact that many polymer reactions take place intramolecularly as exemplified by the radical substitution reaction

(1)

$$\text{~~}\underset{\underset{OO}{|}}{\overset{\overset{CH_3}{|}}{C}}-CH_2-\underset{\underset{H}{|}}{\overset{\overset{CH_3}{|}}{C}}\text{~~} \longrightarrow \text{~~}\underset{\underset{OOH}{|}}{\overset{\overset{CH_3}{|}}{C}}-CH_2-\overset{\overset{CH_3}{|}}{\overset{|}{C}}\text{·~~}$$

This is the reaction of kinetic chain propagation in the chain oxidation of poly-α-olefins which leads to free valence migration along the molecule. In 2,4-dimethyl-pentane, a low-molecular analogue of PP, the ratio of the constants of intra- and intermolecular reactions[1] is about two orders of magnitude at 100°C.[11] No intramolecular transfer of a hydrogen atom from carbon to a peroxide radical can take place unless six-membered (or larger) rings are formed in the transition state. Generally speaking, for a reaction involving the abstraction of a hydrogen atom by a free radical to take place, the bond to be broken must be in the axis of and spaced at a distance of 1·5–2 Å from the unpaired electron; in other words, the pre-start state must be on the reaction coordinate.[8] Apparently, the local conformation of the reaction site will determine the probability of the necessary pre-start state and the probability of formation of a transition complex of the optimum structure. Since structural relaxation is slow in polymers and local mobility depends on the local conformation of a segment of the macromolecule,[12] the kinetics of intramolecular reactions may be expected to be influenced by the conformational set. A few examples will be considered later.

The conformation of macromolecules may also affect, although to a lesser extent, the rates of intermolecular reactions. For example, the rate constant of hydrogen atom abstraction by a tert-butoxyl-radical from atactic PP at 45°C is half that for 2,2,4-trimethyl-pentane[13] owing to steric hindrance, the extent of which must depend on the macromolecular conformation.

In a stressed polymer all the structural and physical parameters of the matrix which affect chemical kinetics are changed. In this review attention will be focused on the effect of tensile stress which has been studied most thoroughly.

The point to be noted first of all is that polymers as a medium consisting of long and flexible chain molecules are characterised by elasticity of two types, one being substantially due to entropy and the other to energy. The generation of entropy-induced stresses in a polymer is associated with a deviation of macromolecular conformation from the statistically most probable one.* The energy stresses arise due to valence angle distortions and changes in bond lengths. The difference between the moduli of elasticity for the entropy and energy stresses amounts to two orders of magnitude (Table 1[14]). Therefore, externally applied tensile stresses cause conformational structural rearrangements giving rise to the internal entropy stresses. Only after these rearrangements have come to an end does distortion of valence angles and bonds which leads to the weakening of the bonds in the macromolecular skeleton take place.

* Because of the internal rotational barrier there is virtually no pure entropy elasticity in polymers, just as no ideal gas exists in nature. It is a question of whether it is the entropy- or energy-induced component of elastic forces which is dominant.

TABLE 1
ELASTIC MODULI FOR MACROMOLECULAR DEFORMATIONS
OF VARIOUS KINDS

Deformation	Design modulus (GPa)
Pure tension of C–C bonds	740
Stretching of zigzag chains	180–340
Bending of C–C bonds	80
Retarded rotation of helices	4–7

It will be shown below that the entropy stresses arising in stretched semi-crystalline polymers are 'frozen' in the amorphous phase of the polymer, unless a special heat treatment is undertaken for stress relaxation. The energy stresses arise in oriented specimens to which additional external load is applied, particularly within the plastic region.

2.   THE EFFECT OF TENSILE STRESSES ON STRUCTURAL AND PHYSICAL PARAMETERS OF POLYMER MATRICES

2.1. The Effect of Tensile Stress on Supermolecular Structure

A tensile stress in highly elastic amorphous polymers, such as rubbers, gives rise to large reversible deformations and orientation which relax after relief of the load. The macromolecular axes tend to orient themselves parallel to the stretch axis. Glassy amorphous polymers respond to forced elasticity and an orientation which remains unchanged after the relief of load at temperatures below the glass transition point. In isotropic semicrystalline polymers, tensile stresses cause profound structural rearrangement: the spherulite structure becomes fibrillar. In the microphotographs in Fig. 1 one can observe the deformation of a unit spherulite of isotactic polystyrene stage by stage.[15] The deformation takes place in a stepwise manner involving the formation of a 'microneck' and an interface is clearly seen between the deformed and undeformed parts of the spherulite. Next, after all the spherulite has passed into the neck, the new oriented structure undergoes a uniform deformation, and the microscopic fibrillar structure takes its final shape. Unlike the spherulite structure, the microfibrillar one is markedly anisotropic. In the microfibrils oriented in the direction of stretching, the zones of amorphous and crystalline states alternate, their dimensions being controlled by the conditions of stretching (and

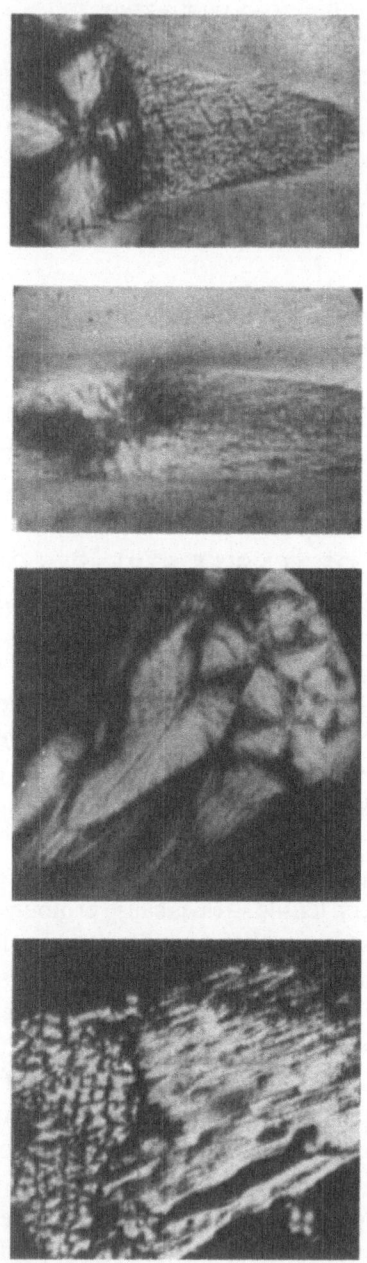

FIG. 1. Stages of spherulite deformation and the neck structure in isotactic PS.

annealing). In the intrafibrillar regions, the individual crystallites are linked up by tie molecules and a network of linkages between the macromolecules, each returning to 'its own' crystallite. Transversely, the individual micro-fibrils are linked to each other by cohesive forces and tie molecules numbering significantly less than in the intrafibrillar regions. Conse-quently, the transverse strength of uniaxially oriented polymers is much lower than the longitudinal strength.

As a result of structural rearrangements due to stresses, redistribution and changes in the solubility of low-molecular compounds and changes in the molecular dynamics of matrices and their conformational set take place in polymers.

2.2. Variation of Solubility of Low-molecular Compounds in Stressed Polymers

As far as elastomers are concerned, this problem has been considered by Kuz'minsky[16] who experimented with rubbers under combined stresses. Static compression stresses bring about an abnormal transport of low-molecular compounds (plasticisers and inhibitors).

The situation is different with tensile stresses: a deformation causing no crystallisation of the rubber (under 200%) has practically no effect on permeability and diffusion;[17,18] larger deformations, leading to crystallis-ation of the rubber, reduce the sorption and diffusion of low-molecular compounds to a considerable extent.[17]

In glassy amorphous polymers and semicrystalline polymers, the effect of tensile stresses on permeability is complex.[19-23] Within the elastic region, the sorption of low-molecular compounds increases with stress as a consequence of an increase in the free volume fraction (the Poisson's ratio of most semicrystalline polymers is < 0·5). In the plastic region the sorption of low-molecular compounds markedly decreases with an increase in stress, because the density of the amorphous component rises in this case.[21] The authors' interpretation of the results has been based on a model of impermeable microfibrils distributed in the bulk of the amorphous inter-fibrillar component. Rough as it is, this model appears to be helpful in most cases owing to the fact that the solubility of low-molecular compounds is proportional to the percentage of the amorphous phase in semicrystalline polymers.[24,25]

As regards oxidation kinetics, the effect of stress on permeability to oxygen is of particular interest. The pertinent data published so far are scarce. It is shown in ref. 25 that the solubility of oxygen in PP is proportional to the percentage of the amorphous phase and depends on the supermolecular structure of the polymer. The point to be noted is that the conception of the impermeability of the crystalline phase for oxygen in semicrystalline polymers is too approximate. For example, all the alkyl radicals originating in the crystalline phase of PP under radiative initiation are transformed into peroxide radicals. In other words, oxygen does diffuse within the crystalline phase of PP (and also within the less dense crystalline phase of poly-4-methylpentene-1). Nothing like that takes place in polyethylene (PE): the dense crystalline regions of this polymer are impermeable to oxygen.[24] This problem is discussed in detail elsewhere (pp. 22–30 of ref. 26). The comparative study, described in ref. 27, of the solubility of oxygen in unannealed (entropy-stressed) and annealed (unstressed) oriented films of isotactic PP has revealed that the solubility in stressed films is 1·5–2·5 times that in annealed ones.

However, the effect of stress on sorption by polymers is not as great as on molecular dynamics and diffusion.

2.3. Molecular Dynamics and Diffusion in Oriented and Stressed Polymers

Deformation does not influence the diffusion characteristics of elastomers strongly when no crystallisation takes place.[17] On the other hand, the diffusion characteristics of semicrystalline polymers may change by as much as several orders of magnitude due to deformation.

In PE the gas diffusion coefficient increases slightly with small reversible deformations. Sorption also increases, apparently, due to an increase in the free volume fraction. But the transition from the region of elastic deformation to that of plastic deformation causes a sharp decrease in the diffusion coefficient (Figs 2(a) and (b)[21]). For example, the three-fold stretching of oriented PE reduces the diffusion coefficient to $\frac{1}{40}-\frac{1}{50}$ of its original value. The decrease is most rapid within a narrow range of stretch ratios apparently coinciding with the formation of the microfibrillar structure.[21] A similar dependence of the diffusion coefficients upon the stretch ratio also exists for nitrogen and methane in oriented PP (Fig. 3).[28]

It will be noted that the effect of polymer deformation on the coefficient of diffusion of low-molecular compounds increases with the size of the diffusate molecule.[20]

Changes in the permeability of polymers after orientational stretching are attributed to the stressed elongated macromolecular fragments present in the amorphous phase. The more relaxed is this phase, the less affected are the coefficients of sorption and diffusion for the same stretch ratios. Therefore, if the temperature of stretching is increased or an oriented polymer is annealed at a temperature higher than that of stretching, the

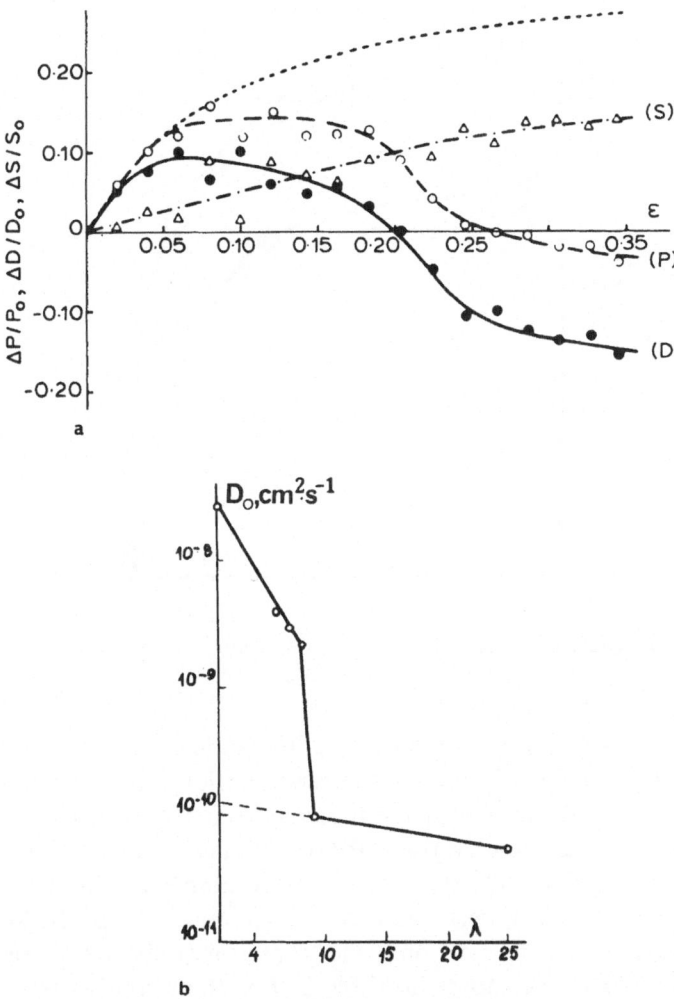

FIG. 2(a).   Relative changes in the sorption, S, diffusion, D, and permeability, P, of CO_2 as a function of the strain ε of isotropic PE film.[21] (b) Changes in the coefficient of CH_2Cl_2 diffusion in PE as a function of the orientational stretching ratio λ.[21]

extent of the change in the permeability characteristics can be markedly reduced.[21]

Apart from molecular dynamics, the coefficient of translational diffusion of low-molecular compounds in oriented stressed polymers is influenced by factors of a purely geometrical nature; in stretched polymer the path of diffusive transfer is longer, since a travelling particle has to circumvent the

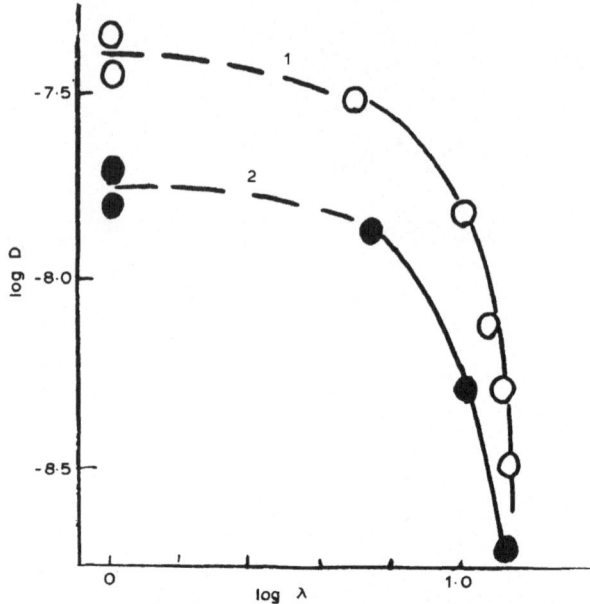

FIG. 3.   The effect of stretch ratio on the coefficients of diffusion of nitrogen (1)
and methane (2) in PP.[28]

impermeable or hardly permeable crystallites (or microfibrils). The effect of
the purely geometrical factors on the coefficient of translational diffusion
can be analysed separately by the paramagnetic probe method (PPM).

The effect of the stretch ratio of PP on the rotational and translational
motion of a stable nitroxyl radical, for example 2,2,6,6-tetramethyl-
piperidine-1-oxyl, has been studied.[29] The relationships between the
stretch ratio, λ, and the rotational and translatory diffusion of the radical
at 80 and 130 °C are shown in Table 2. For the sample with a stretch
ratio $\lambda = 7$, the coefficient of translational diffusion decreases at 80 °C
by a factor of 8 and that of rotational diffusion decreases by a factor of
almost 2 compared with an isotropic polymer. These data have been
obtained with samples into which the radical probe had been introduced
before stretching. However, if the radical is introduced into a prestretched
sample, no changes in the rotational diffusion coefficient, γ_{rot}, compared
with an isotropic polymer are observed at all up to a stretch ratio $\lambda = 7$,
because the radical enters the 'mildest' and most accessible zones of the
sample. Nevertheless, the coefficient of translational diffusion of the radical
decreases compared with the isotropic polymer owing apparently only to
geometrical hindrances which build-up after orientation. From the data

TABLE 2

EFFECT OF STRETCH RATIO OF PP AT 80 AND 130 °C ON THE TRANSLATIONAL AND ROTATIONAL DIFFUSION COEFFICIENTS OF A NITROXYL RADICAL

λ	$80\,^{\circ}C$		$130\,^{\circ}C$	
	$D_{transl}, \times 10^9$ $(cm^2\,s^{-1})$	$D_{rot}, \times 10^{-8}$ (s^{-1})	$D_{transl}, \times 10^7$ $(cm^2\,s^{-1})$	$D_{rot}, \times 10^{-9}$ (s^{-1})
0	5·4	7·9	2·9	7·2
4·5	1·9	6·6	1·7	4·7
7·0	0·7	4·6	1·9	2·8
10	—	4·3	—	2·0

acquired, the coefficient γ, which is a measure of the extension of the path of diffusive transfer due to the geometrical hindrances arising in the course of orientation, has been determined. For samples with stretch ratios $\lambda = 4·5$ and 7, the coefficients are $\gamma = 1·4$ and $\gamma = 2·1$, respectively. Thus, all the changes in the coefficient of translational diffusion in excess of the two-fold changes in specimens with $\lambda = 7$ are due to a decrease in the segmental mobility of macromolecules.

It will be noted that in PP, the radical is significantly hindered and is not very susceptible to stressing at room temperature.[29] Apparently, the low response of the rotational mobility of radicals to external loading of PP at this temperature is attributed to the above phenomenon:[30,31] the correlation time of probe rotation varies by no more than 9% over the stress range 0–250 MPa. At the same time, the effect of the external load on the segmental mobility of oriented PP is quite significant as assessed by the n.m.r. technique.[32]

2.4. The Effect of Tensile Loading on Macromolecular Conformation

Changes in the molecular dynamics of polymers during orientation and loading result from variation of the conformational set: depletion of the mobile, folded conformations and enrichment with the immobile, stretched ones (i.e. a decrease of the entropy of the system[33–39]).

The effect of loading of oriented PE on the concentration of the *gauche* (G) and *trans* (T) isomers is shown in Fig. 4.[38] The effect of the chain stretch ratio on the content of various conformers in the polymer chain has been studied theoretically in ref. 39. In this work it has been shown that at an early stage the chain is stretched chiefly by way of element reorientation: the folded isomers become more regularly distributed along the chain,

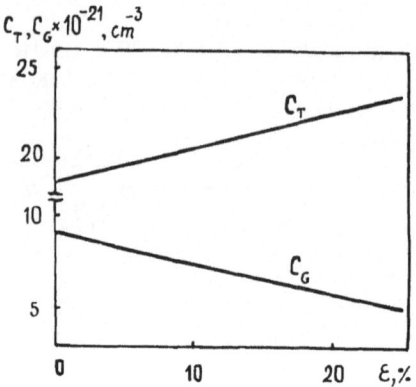

FIG. 4.   The effect of additional deformation along the axis of stretching on the concentration of *trans* and *gauche* conformations in stretched PE ($\lambda = 10$).[38]

preferring the stretched T isomers as their neighbours, and only when the stretch ratio is sufficiently high does the proportion of folded isomers markedly decrease (Table 3). Folded isomers tend to arrange themselves in triads containing one T isomer and two folded ones (Fig. 4). TGT and GTG' are the most stable of the triads containing G isomers in the stretched chain (they are part of the kink of the PE chain), and this is natural: they shorten the chain to a much lesser extent for a given number of G isomers. It has been demonstrated experimentally that stretching of the amorphous interlayers of PE from 40 to 60 Å within the elastic region reduces the number of TGT and GTG' conformations by 30 %, and further stretching tears the specimen.[38] Apparently, the synchronous disappearance of the triads results from the unfolding of the kink, i.e. from the

TABLE 3

THE CHAIN STRETCH RATIO, α, CAUSING A TWOFOLD DECREASE IN THE CONCENTRATION OF G-ISOMERS AND G-ISOMER-CONTAINING DIADS COMPARED WITH THE UNSTRETCHED CHAIN ($\alpha = 0$); χ, THERMODYNAMIC CHAIN STIFFNESS $= E/kT$[39]

χ	α		
	TG	*GG*	*G*
0	0·94	0·68	0·80
1	0·90	0·70	0·81
3	0·86	0·72	0·82

simultaneous transformation of two G isomers into T isomers. This is also verified by the value of the activation energy of deformation (some 8 kcal mol^{-1}) which is twice as high as the potential barrier of the internal rotation in the PE chain.

In ref. 39 it is shown that the disappearance of the TGT and GTG' triads begins at high stretching ratios (~ 0.9). The disappearance of over a third of these conformers when the deformation of PE is close to ultimate indicates that the number of chains stretched to such an extent accounts for more than a third of the total number of tie molecules. The fact that the decrease in the triad concentration takes place in a range of deformation as wide as 50 % makes it possible to assume that some of the fully stretched T chains are further stretched due to distortion of the valence angles.[39]

It should be noted that it was the planar zigzag that was assumed[39] to be the ultimately stretched conformation. This cannot be achieved in PP due to the interaction of the methyl groups and it is therefore the TGT arrangement corresponding to an α-helix in PP crystals which is the ultimate stretched conformation in the polymer.

From the above we conclude that the statistical weight of mobile, folded conformations decreases with the stretching of polymer chains. Pechhold has carried out pertinent calculations for the PE chain.[40] Using his results it was possible to evaluate the contribution of entropy elasticity to the forces tending to contract the stretched polymer chain. The forces in question are of the order of magnitude which is almost the same as that of the contribution of the energy type of elasticity of the kink-isomers, due to the presence of internal rotation barriers, but they have opposite signs.

The effect of the distance between the ends of a partially stretched chain on changes in the free energy F of this chain is plotted in Fig. 5 (p. 129 of ref. 14). The free energy minimum of a chain with n C–C bonds and n_k $2G_1$ kink-isomers corresponds to the distance between the chain ends $r = (n - n_k)\sqrt{2/3}\, a$, where a is the length of the C–C bond. This minimum equals $n_k \Delta U - RT \log Z$, where Z is the number of the ways of locating $n_g/2$ pairs of G and Ḡ linkages among $n/2$ allowed positions. A displacement of the chain ends involving the loss of equilibrium gives rise to elastic forces which are associated with the retarded rotation of the *gauche* bonds outside the zigzag plane in the PE chain and the deformation of the *trans* bonds in the zigzag plane. The modulus, $\bar{E}$, of a chain containing various conformers is given by

$$1/\bar{E} = \sum_i x_i/E_i$$

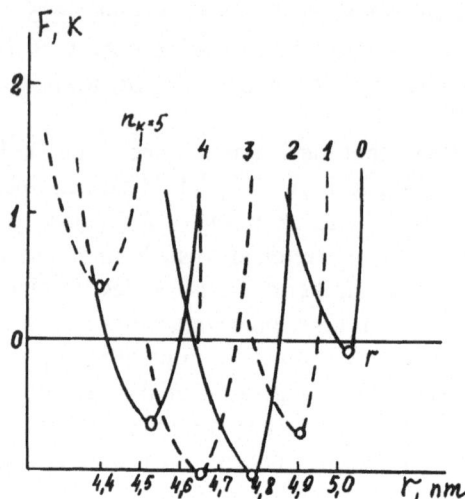

Fig. 5. The effect of the distance between chain ends on the free energy of a partially stretched PE chain (p. 129 of ref. 14).

where E_i are the moduli of individual conformers and x_i are their respective volume fractions.

The less *gauche* conformers present in the chain, the stiffer it is (as few as five kink isomers reduce the modulus of PE chains from 200 to 9·5 GPa). It can be seen from Fig. 5 that the stretching of a chain brings about transitions between the different conformational states; an immediate relaxation of stress is observed in a segment where the conformation becomes more stretched. This means that the process is thermodynamically favourable.

The results of the calculations[39,40] have provided the basis for interpreting the data acquired from studies of the kinetics of oxidation of stressed, oriented polymers.

3.   KINETICS AND MECHANISM OF OXIDATION IN STRESSED POLYMERS

The mechanical activation of oxidation of rubbers discovered by Kuz'minsky *et al.* in 1950,[1,2,41] the well-known effect of the promotion of ozone cracking of stressed rubbers[42–53] and the accelerated breakdown of the C–C bond in the macromolecular skeleton of stressed polymers[54] have created a notion that stresses, as a rule, enhance oxidation. However, this traditional theory was undermined when, beginning in the mid-1970s, a number of studies conducted with various polymeric materials using various methods were published independently and almost simultaneously, providing evidence that the enhancement of chemical reactions, oxidation in particular, by stresses, is far from being a universal phenomenon.[55-68] The point is that, as a rule, oxidation is a complex chain reaction, the various stages of which are affected by stresses in different ways. We will discuss the impact of tensile stress on, firstly, the kinetics of the thermal oxidation of polyolefines with molecular oxygen and ozone and, secondly, the kinetics of the ozonation and oxidation of rubbers.

3.1. Kinetics and Mechanism of Thermal Oxidation of Stressed Polyolefines by Molecular Oxygen

The polyolefine oxidation chain reaction follows the generally recognised scheme given below. It consists of four steps, namely initiation, propagation, degenerative branching and termination of the kinetic chains. Changes in the overall kinetics of the oxidation of stressed polyolefines result from the effect of mechanical stresses on the rates of each of these steps. We will consider the data which are available on the effect of tensile stresses on the rates of each stage separately.

Initiation

(2)

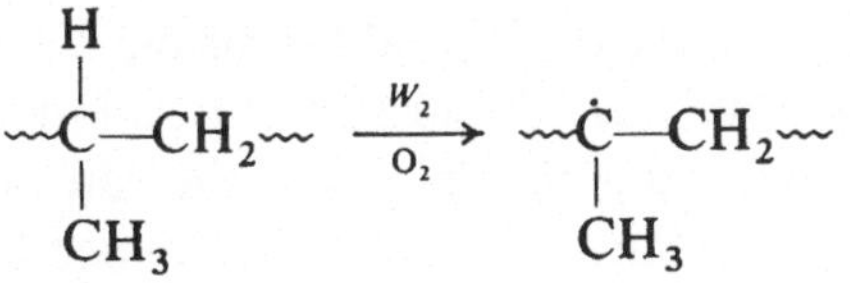

*(2')

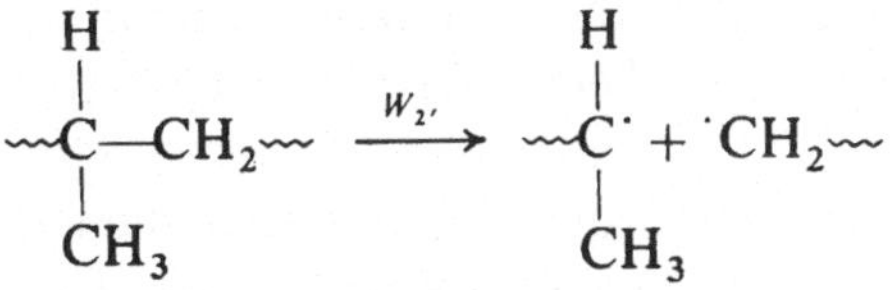

Propagation

$$(3) \qquad \dot{R} + O_2 \xrightarrow{\ k_3\ } R\dot{O}_2$$

*(4)

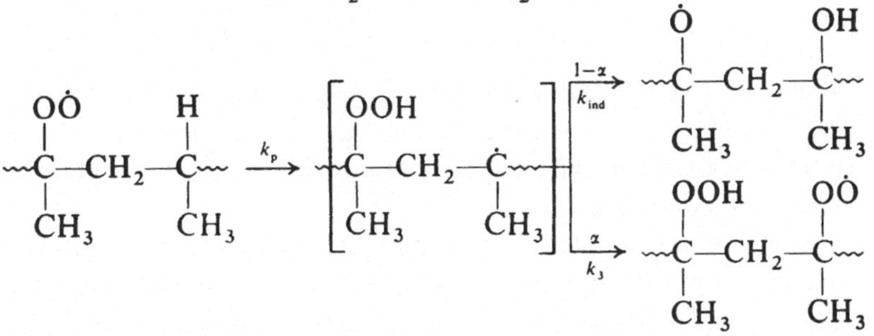

* Asterisks mark the stages at which degradation of the polymer chains occurs.

Branching

(5) $$\text{ROOH} \xrightarrow[(+\text{RH, ROOH})]{k_s} \delta\dot{\text{R}}$$

Termination

(6) $$\dot{\text{R}} + \dot{\text{R}}\rightsquigarrow \xrightarrow{k_6} \left.\begin{array}{r} \\ \end{array}\right\}$$

$$\dot{\text{R}} + \dot{\text{RO}}_2 \xrightarrow{k_{6'}} \left.\right\} \text{ Stable products}$$

$$\dot{\text{RO}}_2 + \dot{\text{RO}}_2 \xrightarrow{k_t} \left.\right\}$$

3.1.1. Initiation

The initiation of the radical chain may follow two different routes in stressed polyolefines. Firstly, by reaction 2, which involves a 'direct' attack of oxygen on macromolecules, leading to the breaking of the C–H bond and formation of non-terminal (tertiary) alkyl macroradicals. This is the 'typical' initiation reaction in unstressed systems which seems to be significantly influenced by various impurities, such as residues from the polymerisation catalyst.[69] Secondly, initiation may proceed from reaction 2′ which involves the breaking of the C–C bonds in the macromolecular skeleton which leads to the formation of terminal macroradicals. This reaction takes place in stressed polymers only.

The effect of structural stresses on the rate of reaction 2, which results from the breaking of the C–H bond by various hydrogen atom acceptors, was studied using unstressed and stressed cyclo-paraffins as model compounds.[62,63] It has been shown[62] that the rate of reaction with ozone increases in stressed eight-, nine- and twelve-membered ring paraffins 48, 89 and 5 times, respectively, compared with an unstressed six-membered ring.[62] The twelve-membered ring paraffin was stressed less than the other materials.

An attempt has been made to interpret these results quantitatively.[64] The rate constant of the reaction with ozone, $k_{i\text{-}m}$, depends on the excess energy available in the ring sustaining tensile stresses, E_s, in the following way:

$$\log(k_{i\text{-}m}/k_{6\text{-}m}) = \gamma(E_s^\varepsilon/(RT)) \qquad \text{(i)}$$

where $k_{6\text{-}m}$ is the rate constant of the reaction with ozone for the six-membered cycle, $k_{i\text{-}m}$ is the same for the i-membered cycle, E_s is the excess energy available in the cycle compared with the six-membered cycle and γ and ε are constants.

It has been suggested that in any reaction involving rehybridisation of a carbon atom from sp^3 to sp^2 and, consequently, an increase in the valence angle, a tensile stres must reduce the activation energy; the increase in the valence angle due to the tensile stress increases the energy of the molecule in its initial state by $E_{init} = 0.5C(\Delta\alpha)^2$, where C is a constant and $\Delta\alpha$ is the deviation of the valence angle from the optimum value.[64] Since the deviation of the valence angle from the optimum value is decreased when a hydrogen atom is abstracted, the value of $E^{\neq}$, the increase in the energy of the transition state due to the tensile stress, will be less than E_{init}, causing a decrease in E_{activ}. This must also hold for the abstraction of a hydrogen atom from a polyolefine molecule.

This hypothesis has been verified experimentally[65] for the ozonation of PP, high-density polyethylene (HDPE) and low-density polyethylene (LDPE). It has been shown that the rate of accumulation of carbonyl groups in PP and PE increases linearly with stress. Oxygen and ozone produce substantially different effects on the rate of stress relaxation under conditions of constant strain. The increase in the difference with stress indicates that the rate of macromolecular degradation by ozone also increases with stress.

An investigation of the surface oxidation of isotropic PP films with different heat treatment histories, using ozone as the oxidant, has shown that the bonds responsible for the long-wave tail of the $973\,cm^{-1}$ band in PP react faster than the other bonds.[66] The authors interpreted this as a result of the more violent reaction between ozone and the most stressed C–H bonds. However, the long-wave branch of the band may correspond not only to overstressed bonds but also to those belonging to other conformations,[70] in which case the above result could be due to the different reactivities of the conformers. Due to lack of information, it is impossible at this time to decide between the alternatives.

The findings in refs 62 and 64–66 were generalised[31] so as to demonstrate the opposite effects of orientation and external loading of stretched specimens on the rate of ozonation of PP and PE. It was shown that neither the changes in the degree of orientation and crystallinity nor in the molecular dynamics of the loaded polymer can offer an explanation. The authors of the above references have come to the conclusion that the reaction of ozone with loaded PP and PE is accelerated mainly by the stress-dependent decrease in the activation energy required for the abstraction of a hydrogen atom from the polymer chain. Unfortunately, the criterion used[31,65,66] for concluding that reaction (2) is accelerated in stressed polymers, is the increasing rate of accumulation of secondary products (carbonyl groups, chain cleavages) resulting from eventual further conversions of the alkyl radicals formed in reaction 2. However, further proof (although also indirect) of the acceleration of reaction 2 in stressed PP, obtained recently,[71] seems to corroborate the point of view of the

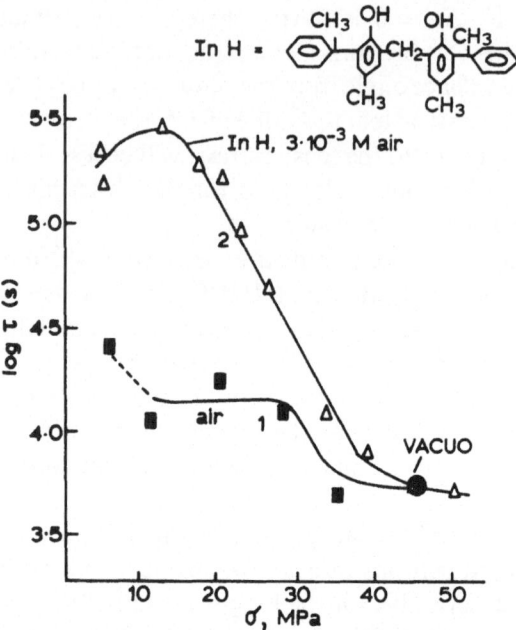

FIG. 6. Isotherms of the durability of oriented PP $(\lambda = 7)$: 1, uninhibited specimen; 2, inhibited specimen. Inhibitor: Nonox; $[InH] = 3 \times 10^{-3} \, mol \, kg^{-1}$. The durability for an inhibited specimen under vacuum is marked with a large circle.[71]

authors of refs 31, 65 and 66 (Fig. 6). There is a certain range of stresses in which the durability, τ_s (the time from the application of stress to the failure of the specimen), of an externally loaded PP in the state of uninhibited oxidation depends on the rate of chain oxidation and is little influenced by the stress[59,72,73] (curve 1 in Fig. 6). In the specimen inhibited with Nonox $(3 \times 10^{-3} \, M)$ chain oxidation was suppressed and, as a result, durability increased at low stresses by more than an order of magnitude compared with the uninhibited specimen. The time over which this occurred was determined by the inhibitor consumption time which, in turn, depends upon the initiation rate. Once the inhibitor has been consumed, oxidative degradation brings about rapid failure of the specimen. Curve 2 in Fig. 6 shows that the durability decreases exponentially as the stress is increased between 15 and 35 MPa, providing evidence that the rate of initiation is increasing. It will be shown below that in this stress range the oxidation is initiated via reaction 2.* Apparently, the real cause of the acceleration of

* Unambiguous interpretation of the results is difficult, since oxidation under load is spatially confined to limited regions (p. 244).

reaction 2 is an extension of the macromolecular segment which occurs due to rehybridisation of the carbon atom from sp^3 to sp^2 in the course of hydrogen atom abstraction and which leads to a partial stress relaxation, rather than the distortion of the valence angles in the initial state because Fig. 6 shows that the absolute values of the applied stresses are quite low.

The second initiation reaction, denoted 2′, and involving breakage of the C–C bond in the macromolecular skeleton, yields terminal macroradicals and takes place in stressed polymers only. Its rate increases exponentially with stress as shown in the numerous publications of the Leningrad school headed by Zhurkov (cf. ref. 55).

3.1.2. Propagation of Kinetic Chains

The first step in the propagation of kinetic chains is the combination of oxygen with the alkyl macroradicals (reaction 3 above). In polymers it is controlled by microdiffusion of oxygen through the polymer matrix.[74–76] The decrease in the translational mobility of gases in PP, which is a result of stretching,[28] must slow down reaction 3 in stretched specimens. This has been demonstrated[77] by a study of the rate of the reaction $\dot{R} + NQ$, a model of the reaction $\dot{R} + O_2$, in isotropic and stretched PP. Figure 7 shows the extent of reaction curves during heating in isotropic (1) and stretched (2) PP specimens at $\lambda = 9$. It can be seen that the decay of alkyl radicals occurs in stretched PP at a slower rate than in isotropic specimens, indicating that the activation energy range has shifted to higher values.

Extra loading of stretched specimens may affect the coefficient of oxygen diffusion, as compared with unstressed polymers, as well as the rate of reaction 3. But there is evidence that within the elastic region these changes are not as obvious as during the transition of a polymer from the isotropic to the oriented state.[20,21] For instance, it may be expected that PP specimens maintained isometric (fixed at the ends), i.e. under elastic strain, will display an oxygen diffusion coefficient at the oxidation temperature (130 °C) which is slightly higher than that in unstretched specimens. However, no experimental data are available so far to support this assumption.

One may expect that in the plastic range the rate of reaction 3 will be even greater. It is known, for example, that alkyl radicals are not transformed into peroxide radicals in an atmosphere of oxygen at 77 K, unless an external load is applied. On the other hand, peroxide radicals are formed in an oxygen atmosphere at 77 K during vibration grinding of various polymers, apparently due to the forced acceleration of oxygen diffusion through the specimen.[78]

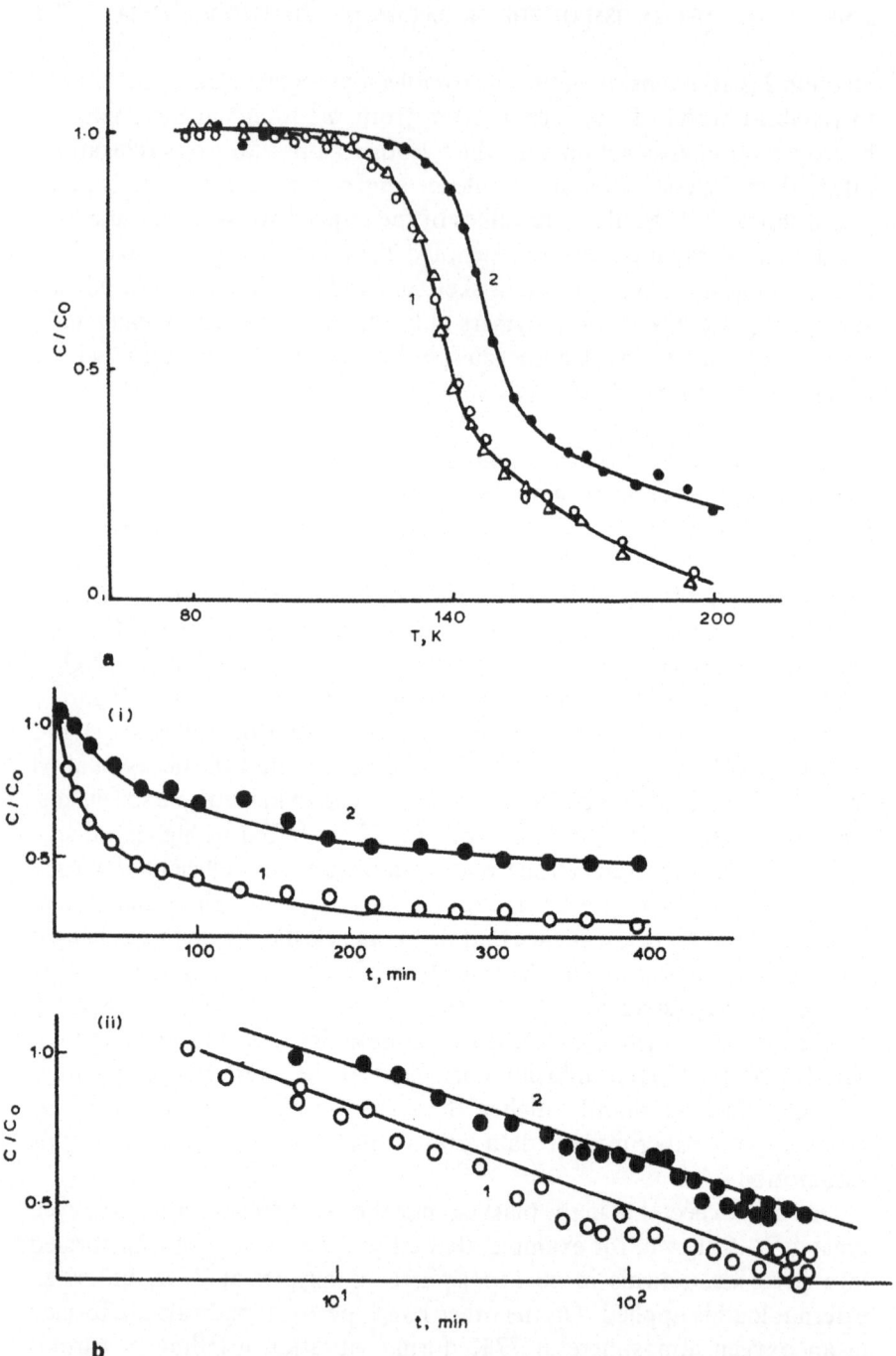

FIG. 7. (a) The heating curves of the reaction $\dot{R} + NO$. 1, isotropic PP specimens; 2, stretched PP specimens ($\lambda = 9$). (b) Kinetic curves of alkyl radical decay in the reaction $\dot{R} + NO$ in PP (i) and the corresponding semilogarithmic plots (ii): notation as in (a).[77]

Thus, it can be concluded that the effect of external loading on the oxygen diffusion coefficient in oriented specimens is not very significant and, apparently, neither is the effect of loading on the rate of reaction 3.

Under sufficiently high oxygen pressures the effect of the rate of reaction 3 on the polyolefine oxidation kinetics manifests itself only indirectly (see below), because it is reaction 4 which controls the propagation rate of the chain oxidation process.

In polyolefines reaction 4 takes place intra- and intermolecularly. For PE and its low-molecular analogues, reaction 4 is, apparently, an intermolecular process. In fact, the constant of hydrogen atom abstraction from a PE macromolecule, with the cumenyl peroxide radical as the acceptor, is $0 \cdot 12$ litres $mol^{-1} s^{-1}$ at $115\,^{\circ}C$,[79] which corresponds to a peroxide radical lifetime in the 'ideal' intermolecular chain transfer in PE in liquid phase of about $2 \cdot 5 \times 10^{-2} s$. One should, of course, take into account the fact that the reactivity of the secondary peroxide radicals in PE is three times higher than that of tertiary cumenyl peroxide and the concentration of the reactive C–H bonds in PE equals $142\, mol\, kg^{-1}$.[80]*

The lifetime of the peroxide radical of PE in intramolecular radical substitution reactions was evaluated while studying $RO_2^{\cdot} \rightarrow R^{\cdot}$ conversions in complexes of PE and urea (PEU) where it is strictly intramolecular.[81] By extrapolating the results thus obtained to a temperature of $115\,^{\circ}C$, it can be deduced that the lifetime of the peroxide radical of PE in the intramolecular reaction 4 is of the order of $3 \times 10^{2} s$, i.e. it exceeds that in the 'ideal' intermolecular reaction by almost four orders of magnitude.

Such a difference may be due to the fact that the probability of formation of an activated complex with an optimum linear structure within a chain segment with the trans-zigzag conformation is low in the PEU complex. A study of the possibility of reaction 4, using a model of the trans-zigzag in PE, has shown that no linear activated complex can be formed, unless the trans-zigzag is distorted either due to a change in the internal rotation angles or distortion of the valence angles. Apparently, it is this fact which causes the activation energy of the intramolecular reaction 4 in PEU to be as high as $24 \cdot 6\, kcal\, mol^{-1}$,[81] in contrast with the activation energy required for hydrogen atom abstraction by the cumenyl peroxide radical in n-decane which is $17 \cdot 5\, kcal\, mol^{-1}$.[80]

The fact that reaction 4 takes place in PE as a predominantly intermolecular process is corroborated by the large amount of isolated (non-hydrogen-bonded) hydroperoxide groups produced in oxidation. The

* In the solid phase the lifetime may be much higher.

oxidation of substituted poly-α-olefines, such as PP and poly(4-methyl-pentene-1) yields mainly hydrogen-bonded hydroperoxide groups but the concentration of the isolated hydroperoxide groups exceeds that of the hydrogen-bonded ones in PE.[82]

To enable intermolecular transfer of the kinetic chains there must be a molecular motion urging the peroxide macroradical to encounter the neighbouring macromolecule. As shown in Sections 2.3 and 2.4, stretching reduces the segmental mobility of macromolecules in the amorphous phase and is likely to affect the rate of reaction 4. In fact, in stretched PE oxidised by radiation in air at room temperature the parameter $k_p/\sqrt{k_t}$ appears to be smaller than in an isotropic specimen by a factor of roughly 1·5.[82] Apparently, this is due to a decrease in k_p, since the decay of radicals is also controlled by molecular dynamics, and in stretched PE the value of k_t must be lower than in an isotropic specimen (see below).

Unlike PE, in substituted poly-α-olefines and their low-molecular weight analogues reaction 4 takes place predominantly at the intramolecular level.[11,83–85] A study of the possibility of an intramolecular reaction 4, carried out for a model of the PP macromolecule, has shown that such a reaction is practically impossible if the radical is confined to the macro-molecular segment with a stretched TGT conformation (identical to the one in PP crystallites), since in this case the peroxide radical is screened from the adjacent C–H bonds of the same macromolecule. Reaction 4 appears to be sterically probable only under conditions of TGT → TGG evolution of the active centre. Formally, the creation of a reaction-promoting conformation may be thought of as an encounter between the radical in the TGT conformation and a G conformer diffusing through the macromolecule. This process is in fact a cooperative transition of the type

$$\text{TGTGTGT} \underset{k_{-r}}{\overset{k_r}{\rightleftharpoons}} \text{TGGTGGT}.$$ The cooperative nature of this transition is

attributed to the fact that the conformations TT and GGG are forbidden in PP (the former due to the 'blocking' of the methyl groups and the latter to the excluded volume effect). It will be noted that the number of G conformers in the right-hand-side of the scheme exceeds that in the left, and conformational transitions of the above type are impossible unless 'surplus' G conformers are available on the macromolecule (as compared to their number in the fully stretched helical TGT sequence). The number of *gauche* conformers within the macromolecular segment is determined by the ratio of the contour length and the end-to-end distance of the segment, and it is controlled 'genetically' for each macromolecule in the amorphous phase of a semicrystalline polymer at the instant of crystallisation or stretching.

Formally, reaction 4 can be considered on the same lines as in describing the kinetics of solid-phase bimolecular reactions.[86,87] Reaction 4 can be described as one in the radical–GG conformer pair:

(7) $$GG + R\dot{O}_2 \underset{k_{-r}}{\overset{k_r}{\rightleftharpoons}} [R\dot{O}_2(GG)] \xrightarrow{k_{ch}} ROOH$$

For $k_r \approx k_{-r}$, we obtain a quasi-stationary approximation of the rate of reaction 4:

$$W_4 = \frac{k_{ch}k_r}{k_{ch} + k_r}[GG][R\dot{O}_2] = k_{effective}[GG][R\dot{O}_2] \qquad (ii)$$

If the encounter between a G conformer and a radical is the limiting step of the reaction, then $k_{effective} \approx k_r$; if the pairwise reaction is the limiting step (i.e. not every 'collision' of a radical with a G conformer leads to reaction), then $k_{effective} \approx k_{ch}$; in an intermediate situation, the values of k_r and k_{ch} are added.

Whatever is the case, the rate of intramolecular reaction 4 is proportional to the concentration of GG conformers on the macromolecule. In the amorphous phase of semicrystalline polymers there is always a variation of the macromolecular segment contour lengths, and consequently, of the concentration of GG conformers.* This implies that oxidation of semicrystalline polymers must involve a distribution of constants k_p while the k_p values (as well as those of k_3 and k_8) as determined from the overall kinetics of oxidation,[88,89] are the effective mean values for a given distribution.

The constant k_4 may be expected to have a minimum value in the crystalline phase of PP, where the non-reactive TGT conformation, which constitutes an 'elementary' link of the helix, prevails. The characteristic reaction time appears to be the time for a helix defect to be formed at the radical localisation point. The effective value of $k_{p(cr)}$ at room temperature, as measured in the study of reaction 4 in the crystalline phase of PP over a temperature range 270–300 K, was about $2 \times 10^{-5}\,s^{-1}$,[90] that is, almost three orders of magnitude smaller than $k_{effective}$ in the amorphous phase of PP.[88,89] Such significant differences between the rates of reaction 4 in the crystalline and amorphous phases apparently results from 'conformational inhibition' of the reaction in the crystalline phase.[60,61]

---

* In an attempt to find out the pattern of distribution of the macromolecular segments depending on their contour lengths, bar charts were plotted representing the concentrations of free radicals after a stepped deformation of PA-6 fibres (p. 195 of ref. 88). Unfortunately, the results obtained were ambiguous, providing no answer to the question as to whether there is a direct correlation between the concentration of the detectable free radicals and that of the cleaved macromolecules.

Evidence of a variation of k_p in PP was obtained[91] when the kinetics of the decay of peroxide radicals, obtained by oxidation of alkyl radicals (created by γ-radiolysis at $-78\,°C$ and room temperature), was investigated. The authors approximated the kinetic curves thus obtained with three and two sets of exponential curves for the above temperatures, respectively. Representation of the results in terms of c/c_0 versus $f(\log t)$ plots gave good straight lines which might be regarded as evidence in favour of the existence of a distribution of the reactive particles according to the constant k_p. Its minimum value has about the same magnitude as the value determined[90] for the crystalline phase of PP.

Stretching and extra loading of polymers brings about a redistribution of conformers (the percentage of GG pairs decreases and that of TG pairs increases) and a decrease in the total concentration of the folded conformers (cf. Section 2.4). As might be expected, this leads to a decrease in the mean effective value of the constant k_p in oriented PP to between one third and one quarter of the value in isotropic PP.[88] Although the changes in the effective constant k_p in oriented PP are small in absolute terms, a shift in the distribution towards lower values for stretched chains enriched with non-reactive GTG conformations is a factor of paramount importance as far as the strength of a material under conditions promoting oxidative degradation is concerned.

3.1.3. Reaction 4 and the Strength of Oxidised PP

The strength of semicrystalline polymers is determined by the 'tail' portion of the size distribution of macromolecular segments in the amorphous phase, which corresponds to the shortest and, therefore, the most stretched chains carrying a minimum number of folded *gauche* conformers. In PA-6 fibres, the fully stretched macromolecules (responsible for the modulus of elasticity and strength of the specimen) and the unstretched macromolecules make up $1 \cdot 1$ and $98 \cdot 6 \%$, respectively[94] (cf. footnote, p. 229). The higher the stretch ratio and the percentage of the short-chain tail of the distribution, the more macromolecules are stretched to the maximum extent and the higher is the initial strength of PP.[58]

It is thus evident that there is the least probability for reaction 4 to take place intramolecularly in such macromolecules. The occurrence of reaction 4 in polyolefines yields hydroperoxide groups which decompose at high temperatures and usually cause the breakdown of the macromolecules, which is the main reason for the decrease in the molecular mass during autooxidation.[92,93] If reaction 4 occurs intramolecularly, a hydroperoxide block structure is formed. In this case, the probability of macromolecular degradation due to hydroperoxide decomposition increases with the length of the block structure according to the relationship,

$$\rho_{bl} = 1 - (1 - \rho_1)^{v_{bl}} \tag{iii}$$

where ρ_1 is the probability of decay of one hydroperoxide group without macromolecular degradation and v_{bl} is the length of the block structure.[93]

Stretched conformers hinder the intramolecular propagation of oxidation chains. The more stretched a segment is, the less GG conformers it has and the lower is the probability that reaction 4 is intramolecular. Therefore, it may be expected that short hydroperoxide block structures or isolated hydroperoxide groups are formed on the stretched macromolecules (in PP, the constant for the intermolecular reaction 4 should be far lower than the intramolecular constant k_4, even in liquid phase, as has been shown for the low-molecular analogues of PP).[11] In a solid polymer the rate of this reaction must be even lower, being limited by the rate of encounter of the reactants, one of which is the stressed polymer chain, and the reaction is hindered by steric, as well as from diffusion factors.[13]

The decay of isolated hydroperoxide groups proceeds at a rate which is a fraction of the decay rate of block structures.[84] This means that both the formation and decay of hydroperoxide groups are hindered on stretched macromolecules. As a result, the rate of oxidative degradation of stretched macromolecules is far smaller than that of folded ones. Since polymer strength is determined by stretched macromolecules, a minimum loss of strength will be observed in those specimens where the concentration of stretched macromolecules is a maximum, i.e. in the highly oriented ones, provided the degree of oxidation is, everywhere, the same. This is demonstrated in Fig. 8.[55]

The amorphous phase of oriented PP formally falls into two unequal parts, a large 'soft' one and a smaller 'stiff' one. In the former, rich in folded conformations, the intramolecular propagation of kinetic chains is a high-rate process yielding hydroperoxide block structures. Degradation processes and the gross kinetics of oxygen absorption are also controlled by this part of the amorphous phase. In the latter, rich in stretched conformations, the intramolecular propagation of kinetic chains is hindered and their oxidative degradation proceeds at a much slower rate. The absorption of oxygen by them has little effect on the gross kinetics of oxidation but oxidised specimens retain their strength only because of these macromolecules. The presence of stretched undegraded macromolecules, even during extensive oxidation of oriented PP, has been shown

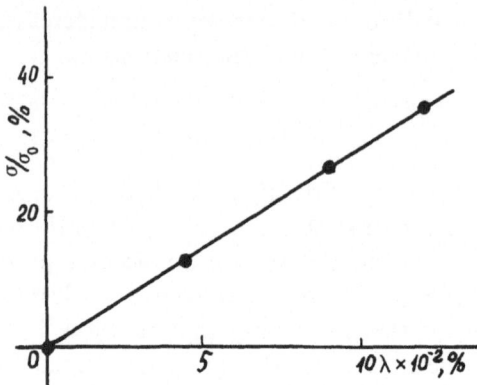

FIG. 8.   Effect of stretch ratio on the relative strength of PP for an absorbed
oxygen concentration of $0.5\,mol\,kg^{-1}$.[55]

experimentally in tests, of annealed and non-annealed PP specimens, for
durability[61] (see Section 3.2.2).

Thus, we come to the conclusion that mechanical stress, which decreases
the diffusive mobility of the polymer matrix and changes the conformational
set of macromolecules in favour of stretched isomers, also slows down the
effective rates of both reactions 3 and 4 of the propagation of oxidation
kinetic chains.

3.1.4. Degenerate Chain Branching

Degenerative branching of kinetic chains takes place in polyolefines as a
result of the thermal decomposition of hydroperoxide groups into radicals.
Since side chain groups are decomposing in this case, stress must have
little effect on the decomposition constant. Nevertheless, the rate of
degenerative branching of kinetic chains is substantially lower in oriented
PP than in the isotropic polymer.[57,95] This is the result of a reduced yield
of hydroperoxide (the branching agent) per mole of the oxygen absorbed.
The kinetic curves of hydroperoxide accumulation in the course of
autooxidation of PP, having different oriented stretch ratios, are given in
Fig. 9. The points corresponding to the absorption of the same concentration of oxygen ($0.5\,mol\,kg^{-1}$) are connected by dashed lines. It can be
seen that the concentration of the accumulated hydroperoxide decreases
with an increase in the stretch ratio of specimens whereas the amount of
oxygen absorbed remains the same. An increase in the rate of thermal
decomposition of hydroperoxide in oriented specimens has no bearing in
this case.[57] The cause of this phenomenon is the decay of the hydro-

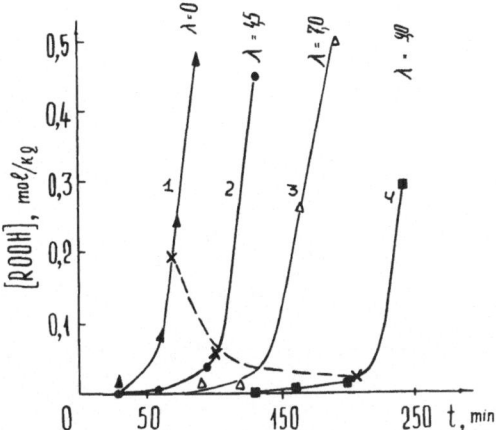

FIG. 9. Kinetic curves of hydroperoxide accumulation due to autooxidation of PP having various stretch ratios (130 °C; oxygen pressure, 600 Torr); the points corresponding to absorbed oxygen concentrations of $0.5 \, \text{mol kg}^{-1}$ are connected by the dashed line. 1, $\lambda = 0$; 2, $\lambda = 4.5$; 3, $\lambda = 7$; 4, $\lambda = 9$.[27]

peroxide group induced by an alkyl macroradical in the β-position, i.e. the isomerisation of a β-hydroperoxide alkyl radical (see the sequence for reaction 4 on p. 221). Competing with the isomerisation reaction is the addition of oxygen to the alkyl macroradical, the product of which is the 'stable' hydroperoxide group. The yield of hydroperoxide per mole of oxygen absorbed is determined within the framework of this mechanism by the expression,

$$\alpha = (k_3[O_2])/(k_3[O_2] + k_{ind}) \qquad \text{(iv)}$$

that is, it varies with the rate of reaction 3.

It is at this stage that the decreased rate of reaction 3, due to the stretching of PP, manifests itself. The reaction of induced hydroperoxide decomposition begins to compete successfully with that of oxygen addition to alkyl macroradical, bringing about a reduction in the yield, α, of hydroperoxide per mole of absorbed oxygen. Consequently, in this case, the main product of reaction 4 of kinetic chain propagation will be an alcohol hydroxyl and alkoxyl macroradical, rather than the branching agent, hydroperoxide, and alkyl macroradical. The effect of oxygen pressure on α supports the above interpretation.[96-98]

A decrease in the hydroperoxide yield per mole of oxygen absorbed in oriented PP specimens is likely to extend the induction period of autooxidation, a fact that has been observed experimentally[58] (see below).

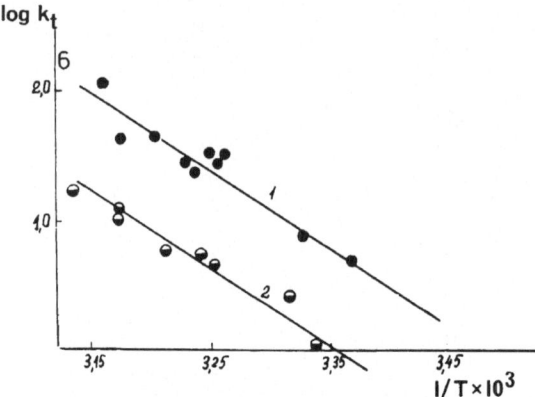

FIG. 10.   Effect of temperature on the decay of peroxide radicals in PP. 1, isotropic
specimen; 2, stretched specimen, $\lambda = 8$.[88]

3.1.5. Termination of Kinetic Chains

In solid polymers the decay of macroradicals is controlled by the rate of
their encounter with each other and is influenced by the intensity of
molecular motion.[8] To account for 1. the high rates of macroradical
decay, 2. the effect of the molecular mobility of a system on the rate of
decay, 3. the unusually high activation energies of decay and 4. the effect of
compensation, all of which were observed experimentally, a number of
models have been proposed and have been critically reviewed in ref. 8.

The rate of peroxide radical decay in the amorphous phase of isotropic
PP ($k_t \sim 10$ litres mol^{-1})[88,99] exceeds that in the crystalline phase by
two or three orders of magnitude even at room temperature[100] (k_t is,
undoubtedly, the effective factor and the activation energy E_t^{PP} equals
28·5 kcal mol^{-1}). Stretching, which decreases the mean segmental mobility
in the amorphous phase, slows down the radical decay rate.[88,101,102] For
PP, the relevant data are given in Fig. 10,[88] which refers to the rate of
radical decay in an elastically-strained entropy-stressed PP polymer. The
relationship between the rate of radical decay and stress within the plastic
region may be quite different. In this case, the intensive mixing of
macromolecules under the influence of a mechanical stress increases the
macroradical encounter and decay rates.[103-105] Radical decay in the
crystalline phase of PE and PP fibres accompanying crystallite deformation
has been observed*[104] (Fig. 11).

---

* This is most likely associated with the generation and migration of defects at the
instant of deformation.

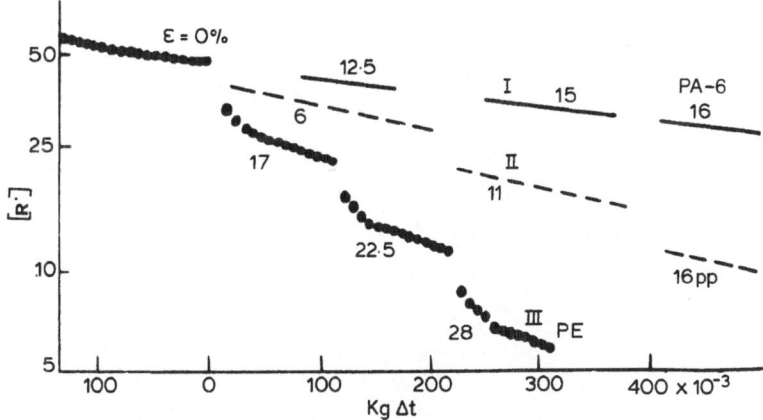

FIG. 11.   Effect of polymer deformation on decrease in the number of radicals in irradiated and heat-treated polymers as a function of time. I, PA-6 fibre; II, PP; III, PE (p. 240 of ref. 14).

An essentially similar picture was also obtained[105] where it was shown that the decay of peroxide radicals, which stops at temperatures below T_g, can be renewed by setting up a short-term plastic deformation in the polymer at 77 K. A reshuffle of macroradicals and macromolecules changes the local surroundings and restores the original pattern of macroradical distribution according to reactivity. Thus, plastic deformation of a polymer must promote macroradical decay. In PP the effect of plastic deformation may be enhanced, because in this case (as will be shown below) the initiation of the kinetic chains predominantly takes the route of reaction 2' yielding terminal radicals which are more mobile and reactive than the radicals in the middle of the chain.

3.2. Effect of Tensile Stress on the Overall Rate of Polyolefine Oxidation
It is convenient to separate the problem into two parts. First the effect of stretching will be considered and second the effect of external loading on the kinetics of polyolefine oxidation. The durability and mechanism of failure of polyolefines will be considered in connection with the second part of the problem.

3.2.1. Effect of Stretching on the Kinetics of Polyolefine Oxidation
This problem has received extensive treatment[26,55-58,60,82,95,98,106,107] and its practical implications have been outlined in a special survey.[26] Here, the main results will be considered briefly.

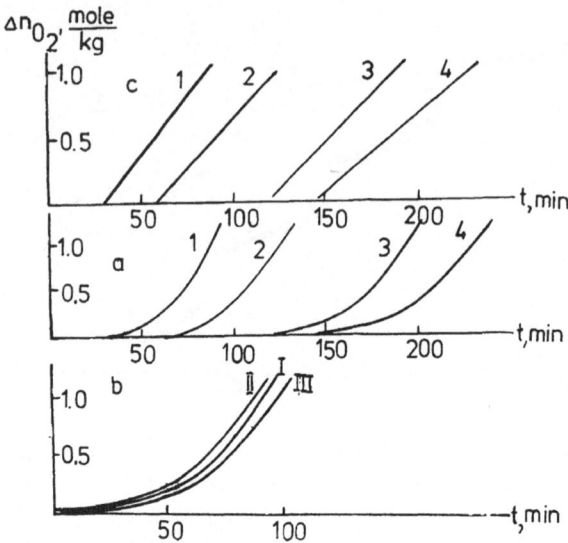

FIG. 12. Kinetic curves of oxygen absorption in PP autooxidation. (a) Un-annealed specimens; (b) annealed specimens; (c) linear representation of the kinetic curves: 1, $\lambda = 0$; 2, $\lambda = 4.5$; 3, $\lambda = 9$; 4, $\lambda = 12$. (130°C; oxygen pressure, 600 Torr).[58]

The kinetics of oxygen absorption during the autooxidation (130°C; oxygen pressure, 600 Torr) of non-annealed and annealed PP films having various stretch ratios are plotted in Fig. 12.[58] A linear representation of the kinetic curves of Fig. 12(a) is given in parabolic coordinates in Fig. 12(c). The relationships between the kinetic parameter of autooxidation and the structural and mechanical properties of stretched specimens is shown in Table 4.

The main features of the oxidation kinetics of stretched PP are as follows:

1. The induction period for the oxidation of unannealed PP specimens increases exponentially with stretch ratio (Fig. 12(a)). A similar relationship was noted for polyamide fibres.[106]
2. There is a difference in the initial oxidation rates during the autocatalytic stage because the slopes of the linear curves in Fig. 12(c) decrease with increasing stretch ratio.
3. The stretch ratio has no effect on the maximum rate of oxidation.
4. The oxygen absorption kinetic curves are approximately parabolic after a certain period of oxidation. In the linear representation, the kinetic curves do not converge at the origin. The time, t_0, obtained

TABLE 4

RELATIONSHIP BETWEEN THE KINETIC PARAMETERS OF AUTOOXIDATION ($130°C$, OXYGEN PRESSURE 600 TORR) AND THE PHYSICAL AND MECHANICAL PROPERTIES OF STRETCHED PP SPECIMENS (UNANNEALED/ANNEALED)

λ^a	$\tau_{corr}^b \times 10^{10}$ ($s, 60°C$)	α_{cr}^c ($\%$)	σ^d (MPa)	E^e (MPa)	τ_{ind}^f (min)	$b,^g \times 10^4$ $((mol\,kg^{-1})^{1/2}\,s^{-1})$	$W_{max},^h \times 10^4$ ($mol\,kg^{-1}\,s^{-1}$)
0	5·4/5·4	58/70	0/—	2	40/30	2·8/2·1	6·9/4·9
4·5	5·7/5·4	66/74	13/—	40	75/30	2·4/—	6·2/4·5
9·0	7·9/5·4	64/76	86/—	290	140/35	2·1/—	6·2/4·5
12·0	9·3/5·4	65/71	215/—	660	170/40	1·9/—	6·2/5·0

^{a} Stretch ratio. ^{b} Rotational correlation time of the spin probe. ^{c} Crystallinity. ^{d} Tensile stress. ^{e} Sample stiffness

$$E = \frac{\partial \lambda}{\partial \sigma}(2\lambda + 1)$$

^{f} Induction period of oxidation. ^{g} The slope of the linear anamorphoses of the kinetic curves in parabolic coordinates. ^{h} Maximum oxidation rate.

by extrapolating the linear representation of Fig. 12(c) on the x-axis has been termed 'the true induction period' of the non-inhibited oxidation of PP.[107] The rate of polymer oxidation is very low during the period t_0.

5. Annealing reduces the differences between stretched and isotropic specimens.

The above data have led to the following basic conclusions:

1. A decrease in the initial rates of oxidation (increase in the induction period) resulting from stretching has, contrary to widespread opinion,[108] no relationship with the degree of crystallinity; annealing markedly increases the degree of crystallinity in stretched PP (from 64 to 76%), whereas the induction period of oxidation becomes much shorter.

2. The variations of the oxidation kinetics are due to the stressed oriented macromolecules present in the amorphous phase; annealing brings about a relaxation of internal stresses and restores the compact macromolecular conformations,[58,109] thereby eliminating the difference between oriented and isotropic specimens. The result of chemical relaxation of internal stresses involving the cleavage of stretched macromolecules during the oxidation process itself is similar (this explains the equality of the maximum oxidation rates).

Thus, the main effect of orientation stretching on PP oxidation kinetics is a substantial increase in the induction period of autooxidation. A similar result is obtained in the oxidation of oriented poly-4-methylpentene-1.[110] In the autooxidation of PE at 130 °C there were no changes in the kinetics of oxygen absorption by isotropic and oriented specimens.[82] This is quite natural, considering that intensive relaxation processes take place in PE at this temperature.

The data given in Section 3.1, for the changes in the chain oxidation reaction kinetics in stressed PP at various stages, indicate that the effective constants k_p and k_t of the rates of propagation and termination, respectively, decrease but in such a way as to cause an insignificant decrease in the parameter $k_p/\sqrt{k_t}$ which determines the overall rate of developed oxidation (by some 33% in an isotropic preoxidised specimen stretched to $\lambda = 4.5$). The initial rate of reaction varies more strongly than the parameter $k_p/\sqrt{k_t}$. For example, it can be seen from Table 4 that the induction period of oxidation increases for a specimen with $\lambda = 12$ more than four-fold compared with an isotropic specimen.

The nature of the effect of stretching of PP on the induction period of oxidation has been investigated.[95,107] Before discussing the results of these studies, it is appropriate to say a few words concerning the reason why an induction period should exist at all in the non-inhibited oxidation of PP and poly(4-methyl-pentene-1) or, in other words, explain the cause of the difference between the kinetics of the early and later stages of oxidation. It is technically difficult to obtain kinetic data at an early stage of the reaction due to the low rates and degrees of oxidation. This lack of information prevents an unambiguous answer to the question raised. It has been shown[107] that kinetics typical of the degenerative branching and quadratic termination reactions cannot become established unless the hydroperoxide concentration builds up to a certain level ($2 \times 10^{-2} \, mol \, kg^{-1}$ of the amorphous phase at 80 °C and an oxygen pressure of 760 Torr) in isotropic and oriented specimens alike (Fig. 13). It is believed that less concentrated hydroperoxide lacks adequate initiating capacity and cannot serve as a basic source of radicals in PP; the initiation then takes place as a direct interaction between oxygen and the macromolecules (reaction 2 on p. 221). Recent findings obtained from studies of the kinetics of PP autooxidation at elevated oxygen pressures corroborate this.[111] It appears that an increase in oxygen pressure over the entire range between 0·02 and 4·24 MPa shortens the induction period, whereas the maximum rate of accumulation of carbonyl groups is practically unaffected by pressure beginning from about 0·45 MPa. The activation energy of the induction

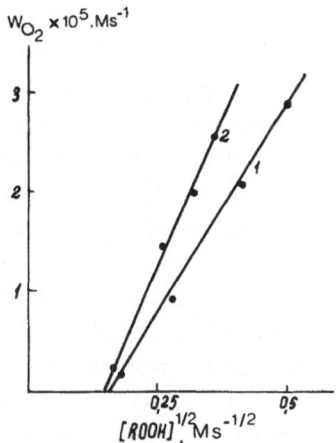

FIG. 13.   Effect of hydroperoxide concentration on the rate of oxygen absorption for an isotropic (1) and stretched ($\lambda = 9$) (2) **PP** film (80°C; oxygen pressure, 760 Torr).[107]

period is $24 \, \text{kcal} \, \text{mol}^{-1}$ at 60–90°C. The reason for the low initiating capacity of hydroperoxide in the initial stage of the reaction may be twofold. The hydroperoxide may decompose either via the monomolecular reaction

$$\text{ROOH} \xrightarrow{k_s''} \text{R}\dot{\text{O}} + \dot{\text{O}}\text{H}$$

(activation energy, $35 \, \text{kcal} \, \text{mol}^{-1}$; low k_p) or via the bimolecular reaction

$$\text{ROOH} + \text{RH} \xrightarrow{k_s'} [\text{R}\dot{\text{O}} + \dot{\text{R}} + \text{H}_2\text{O}] \longrightarrow \delta\dot{\text{R}}$$

yielding rapidly recombining $\text{R}\dot{\text{O}}$ and $\dot{\text{R}}$ radicals (the probability of escape from the cage, δ, is low). At higher concentrations, hydroperoxide decomposition apparently occurs in the intermolecular pairs linked by the hydrogen bond:

$$\text{ROOH} + \text{ROOH} \xrightarrow{k_s} \text{R}\dot{\text{O}}_2 + \text{R}\dot{\text{O}} + \text{H}_2\text{O}$$

This reaction assures a maximum rate of initiation.[85] The initiation constant also increases by roughly an order of magnitude with an increase in the peroxide concentration in the liquid phase, for example, at an early stage of the oxidation of n-decane, during the dissociation of tert.-butylhydroperoxide.[112]

The existence of an induction period during the non-inhibited PP oxidation has been given an alternative explanation.[113] It has been shown, using the ESR technique, that an increase in the hydroperoxide concentration decreases the rate of peroxide radical decay in preoxidised

PP.[114,115] Thus, an increase in the hydroperoxide concentration from 2.5×10^{-3} to 1×10^{-1} mol kg^{-1} caused the value of k_t to decrease from 170 to 5.2 kg mol^{-1} s^{-1} at room temperature, the rate of decrease becoming low after the concentration built up to 4×10^{-2} mol kg^{-1}. Assuming that the constant k_p does not alter with the progress of oxidation, a change in the value of k_t brings about an increase in the parameter $k_p/\sqrt{k_t}$ and the degree of oxidation. This provides, in principle, an explanation for the difference in the kinetics of the initial and later stages of the reaction. However, an induction period of PP oxidation also exists at high temperatures (130 °C and upwards) when the value of k_8 is close to the chemical constant of tertiary radical decay, even in an oxidised polymer, and a further growth in k_8 in the initial stage is not to be expected. Therefore, the hypothesis that the hydroperoxide formed at an early stage in the reaction displays a low initiating capacity is a more plausible one. Anyhow, the established kinetics typical of a developed reaction are associated with the accumulation of hydroperoxide at a certain concentration which is practically the same in isotropic and stretched specimens (Fig. 13).

It is shown in Section 3.1 that the rate of the hydroperoxide-producing reaction decreases in stretched PP as a result of a decrease in the yield of hydroperoxide per mole of the oxygen absorbed. Apparently, this fact is the main cause of the extension of the induction period of oxidation of PP after stretching.[95,107] A reduced hydroperoxide yield, in turn, is the outcome of structural and physical changes occurring during orientational stretching which leads to a slowing down of the rate of microdiffusion of oxygen through the polymer. The consequence is an increase in the rate of induced hydroperoxide decomposition taking place without branching of the kinetic chains. It must be underlined once more that the rate of oxidation of the specimens under an entropy-induced stress is lower than that of unstressed specimens. Stress relaxation and the restoration of the compact macromolecular conformations in the amorphous phase of oriented specimens eliminate the structural and physical differences between the isotropic and oriented specimens and increase the rate of oxidation of the latter (see Table 4). This involves far-reaching implications for the oxidation kinetics and durability of the specimens subjected to external tensile loading.

3.2.2. Oxidation Kinetics, Durability and Failure Mechanism of Polyolefines Under External Tensile Loading

Since stress brings about changes in the structure and molecular dynamics of polymer matrices (see Section 2), it is reasonable to suppose that the kinetics of polymer oxidation under an external load and of unstressed specimens should be quite different. This problem has been considered in a number of studies involving oriented PP specimens.[59,72,73,116] The specimens are exposed to various loads causing their failure under conditions of intensive high-temperature oxidative degradation, and the durability then correlated with the concentration and distribution of the oxidation products in the specimens at failure. It appears that there is a range of stresses over which the load has little effect on the durability of oxidised PP[59] and may even increase with load,[117] being, however, lower than the durability in vacuum by many orders of magnitude. In other words, durability is determined by the rate of oxidative degradation.[71-73] This is exemplified in Fig. 14 by the isotherms of durability and the steady-state creep rate curves versus stress in oxygen and under vacuum for various PP specimens.[73] The durability isotherms can be nominally divided into three segments. In the first, the durability is determined by the rate of the oxidative degradation chain reaction. Within the second segment, the durability is limited by the sum of the rates of stress-induced and oxidative degradation, and in the third segment only by the rate of the stress-induced degradation of the macromolecules subjected to the load. This rate decreases within a certain range of increasing stress because of the orientational strengthening of the polymer. Accordingly, the lifetime significantly increases within the first segment after the introduction of an inhibitor,[60,71] and remains unaltered within the third segment (Fig. 6).

A study of the effect of external load on concentration and the distribution of the oxidation products in PP specimens has revealed new data.[60,116] Tests have been conducted on over 100 specimens with different stretch ratios, initial structure (annealed and unannealed) and stress ($\sigma = 5-70$ MPa). The distribution of the products of oxidation over the film is determined by scanning with an infrared beam focused on an area of 1 mm^2. A summary of the main results is given below.

1. The transition from elastic strain to creep due to an increase in stress involves a change in the concentration and composition of the oxidation products in the zone of specimen failure, which indicates that the reaction initiation mechanism is changing as well: the initiation by reaction 2 gradually turns into that by reaction 2' (see p. 221). Simultaneously, the oxidation chains grow shorter and the oxidation itself is no longer a chain reaction, provided the stress is sufficiently high.

It can be seen from the infrared spectra of the specimens in the region of the carbonyl and hydroxyl group vibration (Fig. 15)[60,116] that the carbonyl

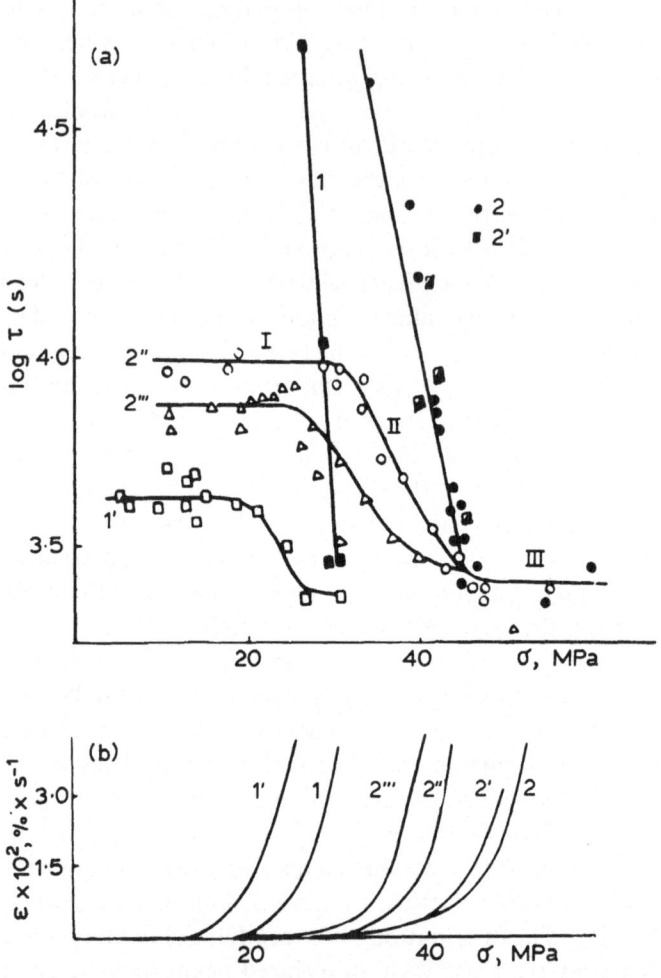

FIG. 14. The durability isotherms (a) and steady-state creep rate versus stress curves (b) at 130 °C. (1, 1') $\lambda = 5$; (2, 2', 2") $\lambda = 7$; (1', 2", 2''') in oxygen; (1, 2, 2') in vacuum; (1, 1', 2, 2') unannealed specimens; (2', 2") annealed specimens.[73]

band in the PP specimens stressed within the elastic region has an identical shape to that of the band in an unstressed oxidised polymer; the maximum of the band at $1710\,cm^{-1}$ corresponds to the valence vibrations of the $C{=}O$ bond in ketones (Fig. 15(a), spectrum 1). An increase in the stress causes the maximum to shift towards higher frequencies, and within the plastic region the maximum is observed at $1740\,cm^{-1}$, indicating that aldehyde groups are the main product (Fig. 15(a), spectrum 3). The

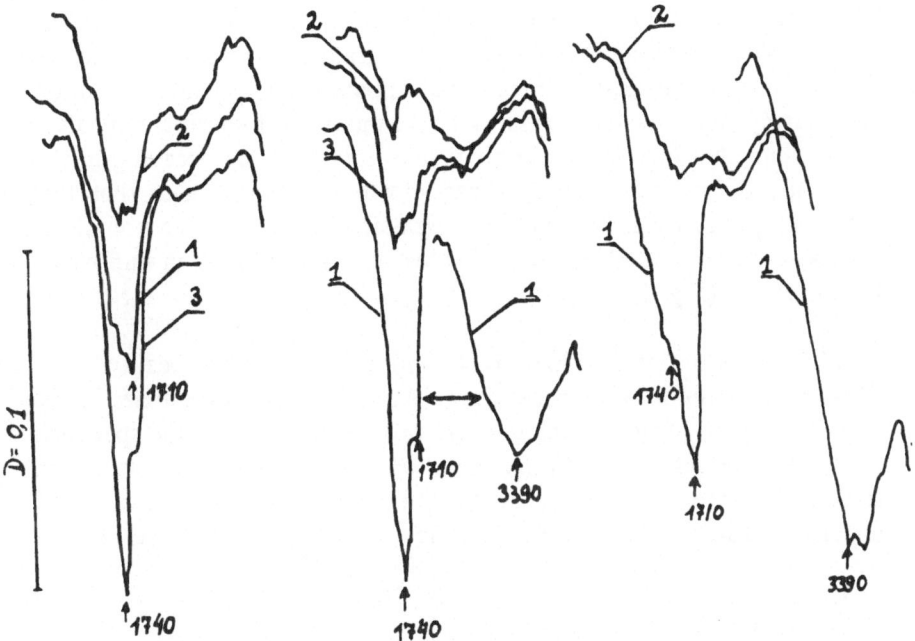

FIG. 15.   Infrared spectra of PP specimens, oxidised under stress, after failure.
(a) Annealed specimens, $\lambda = 7$: (1) $\sigma = 10$ MPa; (2) $\sigma = 13$ MPa; (3) $\sigma = 20$ MPa.
(b) Same specimens: (1) $\sigma = 20$ MPa, zone of failure; (2, 3) matrix. (c) $\lambda = 4$,
unannealed specimens: (1) rupture zone; (2) matrix.[60]

changes in the spectrum of carbonyl groups brought about by changes in
the magnitude of deformation are accompanied by a drastic decrease
in the concentration of hydroxyl groups, including hydroperoxides. When
oxidation takes place in the creep region, it is characterised by the presence
of aldehyde groups which are practically the only carbonyl-containing
oxidation products; then, the hydroperoxide content of the specimens
cannot be determined by means of iodometry and the aldehyde:hydroxyl
concentration ratio amounts roughly to 1:1, that is, it corresponds to the
relation which is to be expected as a result of the disproportionation of
the peroxide radicals, one of which should be a terminal one. This implies
that, within the plastic region, the oxidation reaction becomes a non-chain
process and consists of the formation, oxidation and decay of the terminal
macroradicals. These results indicate that the mechanism of initiation is
changing: at low stress the initiation takes place in accordance with reaction 2, yielding non-terminal tertiary macroradicals which are converted
to aldehydes, ketones, hydroperoxides and other groups through further

chain transformations. As the stress increases, the contribution of the stress-induced breaking of the C—C bonds gradually increases, leading to the formation of terminal macroradicals. Apparently, the chain termination rate sharply increases in this case, since the mobility of terminal radicals is higher than that of the non-terminal ones and they are decaying in the plastically deforming matrix. It was shown in Section 3.1.5 that this speeds up the process. The fact that high-stress oxidation is a non-chain process is corroborated by the ineffectiveness of inhibitors in improving the durability (Fig. 6).

2. The stress-induced oxidation is confined to a certain locality.[60,116] This fact has important consequences. An oxidation centre arises in the film at the zone of future fracture and propagates through the specimen over a distance determined by the magnitude of the stress and the specimen structure. The result is that a gradient of the concentration of oxygenated groups is formed in the failure zone. Referring to Fig. 15, the lower spectra

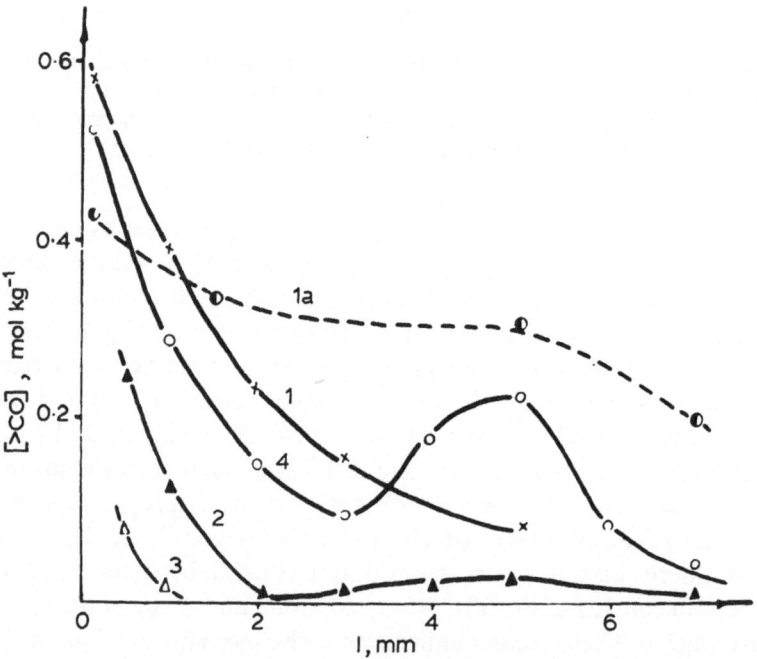

FIG. 16. Concentration profile of carbonyl groups in oxidised PP specimens after failure. Oxidation temperature: (1, 1a) 130 °C; (2–4) 120 °C. (1–3) $\lambda = 8\cdot5$; (4) $\lambda = 4$. (1, 1a) $\sigma = 10$ MPa; (2) $\sigma = 20$ MPa; (3) $\sigma = 29$ MPa; (4) $\sigma = 5$ MPa.[60]

of parts (b) and (c) are those taken in the zone of failure of the specimens and the upper ones refer to matrices (i.e. to localities distant from the zone of failure). For constant stress, the concentration gradient in the unannealed specimens is always higher than in the annealed ones (curves 1 and 1a of Fig. 16). For specimens with a fixed structure, an increase in stress during oxidation causes a sharp increase in the concentration gradient (curves 1–3 of Fig. 16). The result is that the concentration of carbonyl groups averaged per mm^2 of area of the zone of failure sharply decreases with a rise in stress (Fig. 17). By way of illustration, an increase in the stress from 10 to 40 MPa within the elastic region reduces the concentration of carbonyl groups in the fracture zone to as little as 1/20 of the original value. The situation is the same in annealed specimens. Their durabilities differ little from each other within the range of stresses between 10 and 40 MPa, indicating that the mean oxidation rate decreases with an increase in stress in the fracture zone. Since the initiation reaction rate increases with stress (Section 3.1.1) and the radical decay rate decreases within the elastic region (Section 3.1.5), the reduction in the gross rate of oxidation with a rising stress may result from a decrease in the rate of propagation of kinetic chains which is brought about by additional enrichment of the amorphous phase of the polymer with stretched conformations.

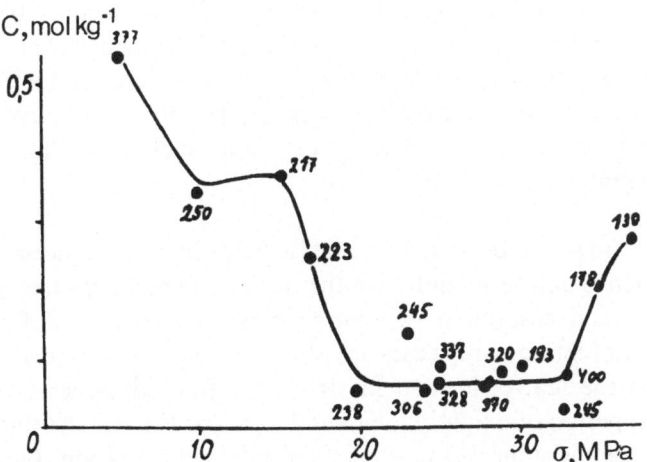

FIG. 17.   Effect of stress on the concentration of carbonyl groups in the rupture zone during oxidation at 120°C; unannealed specimen, $\lambda = 7.5$. The numerals indicate durability.[117]

Within the plastic region, the concentration of the non-chain oxidation products increases with stress (Fig. 17), because the rate of the stress-induced initiation of oxidation increases in this case. A decrease in the rate of propagation of kinetic chains under load and the promotion of their termination within the plastic region are the factors causing the oxidation to become a non-chain process.

3. In addition to concentration the composition of the carbonyl-containing groups may differ in the fracture zone and 'matrix'. Thus, starting from a certain stress level, aldehyde groups may prevail in the matrix whereas in the fracture zone aldehydes and ketones may be present together (Fig. 15(b)). Apparently, this is associated with stress relaxation above and below the folds of the microcracks that foreshadow failure. Thus, the oxidation process which takes place there in the unstressed polymer portions is a chain reaction and yields ketones. Comparing spectra 2 and 3 in Fig. 15(b), it can be seen that the composition of the oxidation products changes from point to point in the matrix, aldehydes alone being identified in spectrum 2 together with ketones in spectrum 3. This result has prompted the conclusion[60] that a stress relaxation zone is set up at the 'point' corresponding to spectrum 3 so that chain oxidation is initiated and ketones are formed. The appearance of stress relaxation zones in a stressed polymer is due to the formation of microcracks. Accelerated oxidation was observed[118] inside the arched microcracks formed during the oxidation of annealed, stretched PP specimens in the diffusive regime. It has been suggested that the difference in the composition and concentration of the oxidation products which exists between the stressed and relaxed zones can find application as the basis for a method of non-destructive searching for defects in stressed oxidised PP,[60] and such a method has been experimentally implemented.[116]

4. The point to be noted is that widely differing concentrations of oxygenated groups accumulate in the matrix of specimens having various structures and sustaining various stresses by the time of failure. For example, not infrequently cases are observed where unannealed specimens fail while the matrix located far from the fracture zone is still in the induction period of oxidation (curves 2 and 3 of Fig. 16). At this time, the fracture zone is subjected to a deep oxidative degradation. On the other hand, the matrix of annealed specimens appears to be deeply oxidised by the time of specimen failure (curve 1a of Fig. 16). The closeness of the durabilities of unannealed and annealed specimens is a striking fact. Even

more striking is the closeness of their repeated durabilities (the time to failure of a fragment of the failed specimen subjected to a repeated loading) in spite of the fact that the matrix of the annealed specimens is, from the outset, oxidised much deeper than that of the unannealed ones.

The following causes may be responsible for this: (a) the strictly local nature of the oxidation of unannealed stressed specimens; (b) the smoothing down of the structural differences between unannealed and annealed specimens around a propagating main crack; (c) the deep oxidative degradation taking place in an annealed polymer which does not affect the stretched and stressed macromolecules which undergo degradation at a lower rate in accordance with the same kinetic law as the stretched macromolecules of unannealed polymer. This is a formidable argument in favour of the suggestions discussed in Section 3.1.2.

5. The marked differences between the concentrations of the oxygenated groups accumulated in the matrix by the time of failure of various specimens show how careful one should be in adopting a macroscopic failure criterion. Very often, a certain concentration of the carbonyl groups accumulated per unit volume of a specimen is used as such a criterion for PP. However, it has been shown that this criterion may give rise to serious errors due to spatial localisation of the oxidation process in stressed PP.[116] It follows from this work[116] that the methods employing averaged values for a specimen should be carefully avoided in studies of the kinetics of failure of stressed polymers.

The spatial localisation of oxidation in stressed polymers and the effect of stress on the reaction kinetics pose new problems for research workers seeking to predict the service lifetime of polymeric products under the combined effect of stress, oxidisers, ultraviolet irradiation, etc. Apparently, the conventional body of mathematics developed for solid-phase reactions is unsuitable for describing the kinetics of localised processes even if a distribution of reaction constants is taken into account. The key question is why and when does the oxidative degradation become spatially localised? Is it localised from the very beginning of the oxidation process in stressed polymer or is this process spread over the entire volume of the specimen for some time? In our opinion, the localisation of oxidation is the aftermath of the opening of microcrack folds and is associated with the acceleration of oxidation in the relaxed zones above and below the folds. Up to that moment the degradation may be considered as taking place over the entire volume of stressed polymer and the formation of a 'hot spot' may be regarded as the result of the accumulation of accidental ruptures. It is possible that the kinetics of this process can be described by the method proposed by Gotlib and Dobrodumov.[119,120]

Briefly the findings in refs 60 and 116 can be summarised as follows:

(1) the initiation mechanism of oxidation varies with stress;
(2) reaction behaviour changes from chain to non-chain oxidation;
(3) establishment of limits for the use of inhibitors to improve the durability of stressed polymers;
(4) the oxidation process in stressed polymers is localised calling for the use of local methods for the experimental investigation of the failure kinetics;
(5) detection of defects in stressed oxidised polymers based on the difference between the concentrations and compositions of the oxidation products formed in the crack zone and in the intact matrix is a possibility.

Two papers devoted to the oxidation kinetics of unstretched, stressed PE[121] and PP[122] specimens appeared in 1982. PE specimens[121] in the form of blades with a width ratio of 3 between the narrow and the wide portions, corresponding to a stress ratio of 3 calculated for the initial cross-sectional area, were oxidised at 70 °C over the range of stresses between 2·9 and 4 MPa and at 90 °C between 1·9 and 3·1 MPa. The oxidation kinetics were monitored by observing the accumulation of carbonyl groups in the wider and narrower portions of the specimens. Under maximum and minimum stresses, the rates of yield of carbonyl groups were the same in the wider and narrower portions. At intermediate stresses the wider specimen portions were always oxidised deeper than the narrower ones, and the wider parts cracked first and caused specimen failure, despite the higher stress sustained by the narrower portion. In other words, those portions of the specimens failed where the deformation was a minimum and the oxidative degradation proceeded at a maximum rate rather than in the most heavily stressed portions. This fully agrees with the findings of other studies.[60,82,116] The lower oxidation rate in the narrower part of the stressed PE specimens is due to stretching and its mechanism is, apparently, the same as that of the decrease in the rate of radiation oxidation in stretched PE.[82]

In ref. 123 it was observed that the rate of photooxidation of isotropic PP decreased under an external load.

Thus, stretching and external loading reduce, as a rule, the overall rate of polyolefine chain oxidation; an increase in the stress converting the reaction to the non-chain form increases the overall rate of oxidation owing to an increase in the rate of the stress-induced initiation of radicals.

3.3. Effect of Stress on the Kinetics of Oxidation of Rubbers with Molecular Oxygen and Ozone

The effect of external loading on the ozone ageing of polyolefines was the subject of several studies,[31,65,66] discussed in Section 3.1.1. In this section some peculiarities of the oxidation of stressed rubbers are considered.

Since most of the elastomers have double bonds, their reactivities are much higher than that of polyolefines. Elastomers are rapidly oxidised in air and even more rapidly in a mixture of ozone and oxygen.

In recent years, this problem has been attracting the ever-increasing attention of scientists in connection with atmospheric pollution and the rising content of aggressive components in the atmosphere.

Mechanical stresses have a significant effect on the oxidation kinetics in elastomers.[124,125] This was pointed out in 1950 by Kuz'minsky[1,2] who termed the phenomenon 'mechanical activation'. 'Mechanical activation' did not involve the cleavage of macromolecules, that is, mechanical degradation, because after repeated straining (amplitude, 75 %; frequency, 250 cpm) the effect was greater by 75 % than in the case of static deformation. However, there was no reason to attribute this to the mechanical activation of oxidation alone, because the dynamic loading promoted, equally, the degradation of vulcanisation in oxygen and in inert atmosphere. The effect was most likely due to microscopic cracking.

It became evident later that stresses may not only activate reactions in stressed elastomers but hinder them as well.[67,126] The authors of these papers attributed activation to a reduction in the activation energy of the oxidation and thermal processes, and retardation of the reaction to a decrease in this value of the pre-exponential factor.

The existing concepts of the mechanism of ozone-induced cracking of stressed elastomers have been critically reviewed by Yu. S. Zuev.[53] In this review we are interested in the effect of mechanical stresses on the kinetics of the oxidative degradation of elastomers.

Ozone ageing of unsaturated vulcanisates involves the breaking of the double bond. The effect of stress on this process was investigated for cyclo-olefines which were used as model compounds,[64] and the results were quantitatively substantiated by computation.[126]

Breaking of the double bond involves a spatial rearrangement of the link

—C1—C2=C3—C4— which is due to transition of the C2 and C3 atoms from the sp^2 to the sp^3 hybrids. The length of the C2—C3 bond increases from 1·333 to 1·534 Å, the valence angle decreases from 122·2 to 112·2°, and the angle of internal rotation decreases from 0 to 60°. This rearrangement is schematically shown in Fig. 18 which also shows the effect of the reaction coordinate, for the *cis* and *trans* fragment —C1—C2=C3—C4—, on the distance between C1 and C4 not bonded by a valence. For the *trans* configuration this distance simply increases in the course of the reaction; for the *cis* configuration it passes through a minimum. This means that the rate of ozonation must be susceptible to stress; it is promoted by a tensile stress if the fragment is of the *trans* configuration and retarded in the case of the *cis* configuration. In the latter case the energy of the transition complex may increase under a tensile stress more than the energy of the initial state so that the activation energy of the ozonation of *cis* rubber must also increase.

This provides a basis for the interpretation of the results of the ozonation of the unsaturated polymer vulcanisates of *cis* and *trans* rubbers.[68,127] In fact, tensile stresses increase the rate of ozonation of *trans*-1,4-poly-

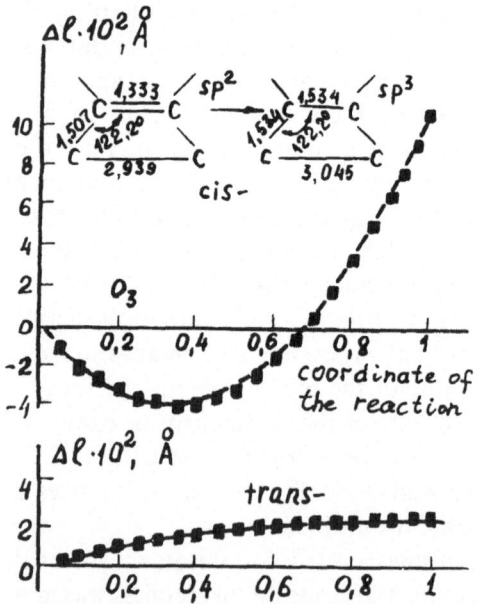

FIG. 18. Effect of the transition of C2 and C3 atoms from the sp^2 to sp^3 hybrid forms on the C1 to C4 distance in the *cis* and *trans* fragments for C1—C2=C3—C4.[68]

butadiene, retard the ozonation of *cis*-1,4-polyisoprene (under a deformation exceeding 20 %) and do not influence the rate of reaction in the case of 1,2-polybutadiene, where the double bond is in the side-chain. Thus a new explanation is given[68,128] of the fact well known, but so far lacking adequate interpretation, that there is a critical deformation during the ozone-induced cracking of stressed rubbers. The decay of oxygenated functional groups accumulated in elastomers during oxidation is activated in statically deformed rubbers.[41]

The information presented in this review obviously indicates that stresses interfere with the process of chain oxidation at all of its stages and are capable of strongly influencing the kinetics of oxidative degradation and even the type of oxidation. Stresses may bring about a sharp spatial localisation of the oxidation process in polymer specimens. They are frequently 'frozen' into the material while it is being processed and revive in the product, affecting its service characteristics. Finally, stresses may arise during the process of ageing if the latter is of a diffusive nature. Uniform oxidation in the volume of specimens seems to be more an exception than a rule in real conditions. This poses new problems for researchers trying to predict the life expectancy of polymer products under actual service conditions, because the use of the method of homogeneous kinetics is inadequate in this case.

APPENDIX

Some results obtained recently seem to throw new light on the origin of primary initiation of PP autoxidation. These involve the investigation of the change in molecular-mass distribution (MMD) in the course of PP autooxidation.[129] It was found that the intensive destruction and the decrease in $\bar{M}_n$ and $\bar{M}_w$ begin only after the completion of the induction period of autooxidation. In the course of the induction period the distribution peak tends to shift towards higher molecular masses. The lower the O_2 pressure during autooxidation, the more is the MMD displaced to higher masses.[129] These findings are more marked in stretched (drawn) PP. Some results are summarised in Table A1; the reproducibility of the results is poor, however, which is obviously the consequence of the extreme sensitivity of the duration of the induction period to polymer structure.

The displacement of the distribution peak to higher masses indicates that crosslinking prevails over chain scission in the induction period.

TABLE A1

CHANGES IN THE MOLECULAR-MASS CHARACTERISTICS OF UNSTRETCHED AND STRETCHED PP DURING AUTOOXIDATION (130 °C; OXYGEN PRESSURE, 600 Torr) AND *In Vacuo* (130 °C)

λ	t (min)	Oxygen				In Vacuo		
		O_2 (mol kg^{-1})	$\bar{M}_n$ ($\times 10^{-4}$)^a	$\bar{M}_w$ ($\times 10^{-5}$)^a	$\bar{M}_z$ ($\times 10^{-6}$)^a	$\bar{M}_n$ ($\times 10^{-4}$)^a	$\bar{M}_w$ ($\times 10^{-5}$)^a	$\bar{M}_z$ ($\times 10^{-6}$)^a
0	0	0	3·6 + 4·0	3·3 + 3·6	1·6 + 2·1	3·6 + 4·0	3·3 + 3·6	1·6 + 2·1
0	5	0	3·4	4·2	2·3	—	—	—
0	18	0·04	5·1	3·5	2·8	4·3	3·4	2·6
0	30	0·05	3·2	2·0	2·2	3·9	3·8	2·8
0	45	0·14	2·2	0·6	0·13	—	—	—
0	60	0·24	3·6	1·4	0·5	4·4	3·0	2·0
0	80	1·4	0·8	0·16	0·03	4·4	3·0	2·0
9	0	0	3·6	2·9	1·6	3·6	2·9	1·6
9	15	0	3·1	2·5	1·6			
9	80	0	3·4	2·6	1·5	—	—	—
9	110	0·14	4·1	6·3	1·5	4·4	2·7	1·0
9	150	0·24	5·1	3·3	2·4	4·7	2·4	—

^{a} The intervals for $\bar{M}_n$, $\bar{M}_w$ and $\bar{M}_z$ values in different experiments are given.

As mentioned above, chain scission is mainly linked to hydroperoxide decomposition. We can conclude that hydroperoxide decomposition does not play an important part in the induction period. The main initiation reaction is then the direct interaction of oxygen molecules with polymer chains; this may proceed as a bimolecular and trimolecular reaction:

$$(8) \quad RH + O_2 \xrightarrow[k_8]{W_8} [R^{\cdot} + HO_2^{\cdot}] \xrightarrow[O_2]{k_3} [RO_2^{\cdot} + HO_2^{\cdot}]$$

$$\xrightarrow{k_{rec}} ROOH \qquad \xrightarrow{k_{dis}} \text{Products}$$

$$\xrightarrow{k_{d_1}} R^{\cdot} + HO_2^{\cdot} \qquad \xrightarrow{k_{d_2}} RO^{\cdot} + HO_2^{\cdot}$$

$$(8') \quad 2RH + O_2 \xrightarrow[k_{8'}]{W_{8'}} [R^{\cdot} + R^{\cdot} + H_2O_2] \qquad \text{(I)}$$

$$\xrightarrow{k_{cross}} R\text{—}R$$

$$k_3 \downarrow O_2$$

$$R^{\cdot} + RO_2^{\cdot} \xleftarrow{k_{d_3}} [RO_2^{\cdot} + R^{\cdot}] \text{ (II)} \xrightarrow{k_{rec}} ROOR$$

$$k_3 \downarrow O_2$$

$$RO_2^{\cdot} + RO_2^{\cdot} \xleftarrow{k_{d_4}} [RO_2^{\cdot} + RO_2^{\cdot}] \text{ (III)} \xrightarrow{k_{dis}} \text{Products}$$

Then for reaction 8', assuming that initiation occurs only from (III):

$$W_i = k_{8'}[RH]^2[O_2] \cdot \left\{ \frac{2k_3[O_2]}{2k_3[O_2] + k_{cross}} \cdot \frac{k_3[O_2]}{k_3[O_2] + k_{rec}} \cdot \frac{k_{d_4}}{k_{d_4} + k_{dis}} \right\} \quad (?)$$

$$W_{cross} = k_{8'}[RH]^2[O_2] \cdot \frac{k_{cross}}{k_{cross} + k_3[O_2]} \quad (?)$$

In reaction 8 one cannot expect crosslinking to occur easily, because high- and low-molecular radicals are created in the cage. In reaction 8' crosslinking is much more probable, because two macroradicals are created in the initial cage (I). Crosslinking proceeds if two alkyl radicals recombine in the cage; if the diffusion of oxygen in the polymer matrix is fast enough, the primary cage may be converted into the secondary one (II), which also has a good chance of decaying by recombination; only the third cage would not give crosslinking, because tertiary peroxide radicals disproportionate rather than recombine.

Thus, the crosslinking and initiation rates would depend on the oxygen permeation parameters (through k_8 and $[O_2]$) and on molecular motion in the polymer matrix (through k_d); the constants k_8 and $k_{8'}$ in the solid phase may also depend on molecular motion. It is shown above that permeation parameters and molecular motion are changed significantly by stretching. The lower probabilities f_1 and f_2, of cage (II) and cage (III) formation in stretched PP, may be one of the main causes of low hydroperoxide yield in stretched samples: oxygen is consumed in reaction 8', but only a few free peroxide radicals are formed (low W_i); in stretched PP the formation of crosslinks would be more preferable. We believe the experimental results confirm this point of view (Table A1).

The same phenomena may be expected in unstretched PP in the case of low oxygen pressure and this is observed.[129]

In the light of these results we believe that the primary initiation reaction in PP is the trimolecular reaction 8'.

REFERENCES

1. KUZ'MINSKY, A. S., MAIZELS, M. G. and LEZHNEV, N. I., *Dokl. AN SSSR*, **71** (1950), 319.
2. KUZ'MINSKY, A. S. and MAIZELS, M. G., *Chemistry and Physical Chemistry of High-Molecular Compounds*, Moscow, Nauka, 1952 (in Russian).
3. JECH, G. S. J., *J. Macromol. Sci.*, **136** (1972), 451.
4. ARZHAKOV, S. A., BAKEEV, N. F. and KABANOV, V. A., *Vysokomol. Soed.*, **15A** (1973), 1154.

5. LEBEDEV, V. P., *Uspekhi Khimii*, **47** (1978), 127.
6. GRIVA, A. P. and DENISOV, E. T., *J. Polym. Sci.*, *Chem. Ed.*, **14** (1976), 1051.
7. RAPOPORT, N. YA., SHLYAPNIKOV, YU. A., GROMOV, B. A. and DUBINSKY, V. ZH., *Vysokomol. Soed.*, **14A** (1972), 1540.
8. EMANUEL, N. M., ROGINSKY, B. A. and BUCHACHENKO, A. L., *Uspekhi Khimii*, **51** (1982), 361.
9. GRIVA, A. P., DENISOVA, L. N. and DENISOV, E. T., *Kinetika i Kataliz.*, **19** (1978), 309.
10. GRIVA, A. P., DENISOVA, L. N. and DENISOV, E. T., *Vysokomol. Soed.*, **21A** (1979), 849.
11. MILL, T. and MONTORSI, G., *Int. J. Chem. Kinetics*, **5** (1973), 119.
12. ANUFRIEVA, E. V., GOTLIB, YU. YA., KRAKOVYAK, M. G. and SKOROKHODOV, S. S., *Vysokomol. Soed.*, **14A** (1972), 1430.
13. NIKI, E. and KAMIYA, I., *J. Org. Chem.*, **38** (1973), 1403.
14. KAUSH, H., *Polymer Fracture*, Moscow, Mir, 1980, p. 126 (in Russian).
15. KARGIN, V. A., SOGOLOVA, T. I. and RAPOPORT, N. YA., *Vysokomol. Soed.*, **6** (1964), 1562.
16. KUZ'MINSKY, A. S., *Developments in Polymer Stabilization—4*, ed. G. Scott, Applied Science Publishers, London, pp. 71–110.
17. BARRIE, J. A. and PLATT, B., *J. Polym. Sci.*, **49** (1961), 479.
18. IASUDA, H. and PETERLIN, A., *Macromol. Chem.*, **73** (1964), 180.
19. BRANDT, W. W., *J. Polym. Sci.*, **18** (1974), 531.
20. IASUDA, H. and PETERLIN, A., *J. Appl. Sci.*, **18** (1974), 531.
21. PETERLIN, A., *J. Macromolec. Sci.-Phys.*, **11B** (1975), 57.
22. VASENIN, R. M., SHAPKHAEV, E. G. and TCHALYKH, A. E., *Dokl. An SSSR*, **197** (1971), 976.
23. WILLIAMS, J. L. and PETERLIN, A., *J. Polym. Sci.*, *A2*, **9** (1971), 1483.
24. MICHAELS, A. S. and BIXLER, J. F., *J. Polym. Sci.*, **50** (1961), 393.
25. KIRYUSHKIN, S. G., YAKIMCHENKO, O. E., SHLYAPNIKOV, YU. A., PARIYSKY, G. B., TOPTYGIN, D. YA. and LEBEDEV, YA. S., *Vysokomol. Soed.*, **17B** (1975), 385.
26. RAPOPORT, N. YA., *Developments in Science and Engineering, Chemistry and Technology of High-Molecular Compounds*, **13** (1980), 3 (in Russian).
27. RAPOPORT, N. YA., BERULAVA, S. I., KOVARSKY, A. L., MUSAELYAN, I. N., ERSHOV, YU. A. and MILLER, V. B., *Vysokomol. Soed.*, **17A** (1975), 2521.
28. TOCHIN, V. A., SAPOZHNIKOV, D. N., SHLYAKHOV, R. A., MASALIMOV, K. T. and MUSAELYAN, I. N., Diffusion effects in polymers, *Theses of Reports at 3rd All-Union Conference*, Riga, 1977, p. 156 (in Russian).
29. BARASHKOVA, I. I., VASSERMAN, A. M. and RAPOPORT, N. YA., *Vysokomol. Soed.*, **21A** (1979), 1683.
30. POPOV, A. A. and KARPOVA, S. G., *Vysokomol. Soed.*, **22A** (1980), 868.
31. POPOV, A. A., KRISYUK, B. E., BLINOV, N. N. and ZAIKOV, G. E., *Europ. Pol.J.*, **17** (1981), 169.
32. EGOROV, E. A. and ZHIZHENKOV, V. V., *Fiz. Tverd. Tela*, **8** (1966), 3853; *Vysokomol. Soed.*, **10A** (1968), 451.
33. NIKITIN, V. N., VOLKENSTEIN, M. V. and VOLCHEK, B. Z., *Zh. Tekhn. Fiziki*, **25** (1955), 2486.

34. VOLCHEK, B. Z. and NIKITIN, V. N., *Zh. Tekhn. Fiziki*, **28** (1958), 1753.
35. NOVAK, I. I., VETTEGREN, V. I. and PETROVA, S. P., *Vysokomol. Soed.*, **11B** (1967), 403.
36. GAFUROV, U. G. and NOVAK, I. I., *Mekh. Polym.*, **1** (1970), 171.
37. NOVAK, I. I., SHABLYGIN, M. V., PAKHOMOV, P. M. and KORSUKOV, V. E., *Mekh. Polym.* (1975), 1077.
38. PAKHOMOV, P. M., SHERMATOV, M. L., KORSUKOV, V. B. and KUKSENKO, V. S., *Vysokomol. Soed.*, **18A** (1976), 132.
39. DARINSKY, A. A. and NEELOV, I. M., *Vysokomol. Soed.*, **20A** (1978), 2381.
40. PECHHOLD, W., *Kolloid-Z u. Z-Polymere*, **228** (1968), 1.
41. SOKOLOVSKY, A. A., UKHOVA, E. M., BANDURINA, V. A. and KUZ'MINSKY, A. S., *Vysokomol. Soed.*, **20B** (1978), 142.
42. BARTENEV, G. M. and ZUEV, YU. S., *Strength and Failure of Highly Elastic Materials*, Moscow–Leningrad, Energiya, 1964 (in Russian).
43. ANDREWS, E. H., *J. Appl. Polym. Sci.*, **10** (1966), 47.
44. SALOMON, G. and VAN BLOOIS, F., *J. Appl. Polym. Sci.*, **8** (1964), 1991.
45. DE VRIES, K. L., SIMONSON, E. R. and WILLIAMS, M. L., *J. Macromolec. Sci. Phys.*, **B4** (1970), 671.
46. DE VRIES, K. L., SIMONSON, E. R. and WILLIAMS, M. L., *J. Appl. Polym. Sci.*, **14** (1970), 3049.
47. SEDLACEK, B., OVERBERGER, C. B., MARK, H. F. and FOX, G. (Eds), Degradation and stabilization of polyolefines, *Polymer Symposium*, 1976, p. 57.
48. ZUEV, YU. S., *Failure of Polymers in Aggressive Media*, Moscow, Khimiya, 1972 (in Russian).
49. KIM, C. S., *Rubber Chemistry and Technology*, **42** (1969), 1095.
50. GENT, A. N. and HIRAKAWA, H., *J. Polym. Sci.*, A2, **5** (1967), 157.
51. GENT, A. N. and McGRATH, J. E., *J. Polym. Sci.*, A2, **3** (1965), 1475.
52. BRADEN, M. and GENT, A. M., *J. Appl. Polym. Sci.*, **1** (1960), 90.
53. ZUEV, YU. S., *Failure of Elastomers Under Typical Service Conditions*, Moscow, Khimiya, 1980 (in Russian).
54. REGEL, V. R., SLUTSKER, A. I. and TOMASHEVSKY, E. E., *The Kinetic Aspects of the Strength of Solids*, Moscow, Nauka, 1974 (in Russian).
55. RAPOPORT, N. YA. and MILLER, V. B., *Dokl. AN SSSR*, **227** (1976), 911.
56. RAPOPORT, N. YA., GONIASHVILI, A. SH., AKUTIN, M. S. and MILLER, V. B., *Vysokomol. Soed.*, **19A** (1977), 2211.
57. RAPOPORT, N. YA. and MILLER, V. B., *Vysokomol. Soed.*, **18A** (1976), 2343.
58. RAPOPORT, N. YA., LIVANOVA, N. M. and MILLER, V. B., *Vysokomol. Soed.*, **18A** (1976), 2045.
59. LIVANOVA, N. M., RAPOPORT, N. YA., MILLER, V. B. and MUSAELYAN, I. N., *Vysokomol. Soed.*, **18A** (1976), 2260.
60. RAPOPORT, N. YA., *Dokl. AN SSSR*, **264** (1982), 1436.
61. RAPOPORT, N. YA., LIVANOVA, N. M., GRIGOR'EV, A. G. and ZAIKOV, G. E., *Vysokomol. Soed.*, **25**, in press.
62. POPOV, A. A., RAKOVSKY, S. K., SHOPOV, D. M. and RUBAN, L. V., *Izv. AN SSSR, Ser. Khim.* (1976), 982.
63. FERGUSON, K. and WHITTLE, E., *Trans. Faraday Soc.*, **67** (1971), 2619.
64. POPOV, A. A. and ZAIKOV, G. E., *Dokl. AN SSSR*, **244** (1979), 1178.

65. POPOV, A. A., KRISYUK, B. E., BLINOV, N. N. and ZAIKOV, G. E., Dokl. AN SSSR, 253 (1980), 1169.
66. KRISYUK, B. E., POPOV, A. A. and ZAIKOV, G. E., Vysokomol. Soed., 22A (1980), 329.
67. BOLSHAKOVA, S. I. and KUZ'MINSKY, A. S., Vysokomol. Soed., 21B (1979), 145.
68. POPOV, A. A., BLINOV, N. M., KRISYUK, B. E. and ZAIKOV, G. E., Europ. Polym. J., 18 (1982), 413.
69. BILLINGHAM, N. C., CALVERT, P. D. and KNIGHT, J. B., Proc. IUPAC, 28th Macromolec. Symposium, Amherst, 1982, p. 291.
70. GAFUROV, U. G., Mekh. Polym., 3 (1978), 544.
71. LIVANOVA, N. M., RAPOPORT, N. YA. and ZAIKOV, G. E., Eur. Polym. J., in press.
72. LIVANOVA, N. M., RAPOPORT, N. YA. and MILLER, V. B., Vysokomol. Soed., 20B (1978), 503.
73. LIVANOVA, N. M., RAPOPORT, N. YA. and MILLER, V. B., Vysokomol. Soed., 20A (1978), 1893.
74. YAKIMTCHENKO, O. E., GAPONOVA, I. S., GOLDBERG, V. M., PARIYSKY, G. B., TOPTYGIN, D. YA. and LEBEDEV, YA. S., Izv. AN SSSR, Ser. Khim. (1974), 354.
75. YAKIMTCHENKO, O. E., KIRYUSHKIN, S. G., PARIYSKY, G. B., TOPTYGIN, D. YA., SHLYAPNIKOV, YU. A. and LEBEDEV, YA. S., Izv. AN SSSR, Ser. Khim. (1975), 2235.
76. MIKHAILOV, A. J., KUZINA, S. I., KUKOVNIKOV, A. F. and GOL'DANSKY, V. I., Dokl. AN SSSR, 204 (1972), 282.
77. YAKIMTCHENKO, O. E., ANISONYAN, K. E. and RAPOPORT, N. YA., High-Molec. Comp., 23A (1981), 703 (in Russian).
78. BRESLER, S. E., ZHURKOV, S. N., KAZBEKOV, E. N., SAMINSKY, E. M. and TOMASHEVSKY, E. E., Zh. Tekhnicheskoi Fiziki, 29 (1959), 358.
79. IVANCHENKO, P. I., DENISOV, E. T. and KHARITONOV, V. V., Kinetika i Kataliz, 12 (1971), 12, 492.
80. DENISOV, E. T., MITSKEVICH, N. I. and AGABEKOV, V.-E., The Mechanism of Liquid-Phase Oxidation of Oxygen-Containing Compounds, Minsk, Nauka i Tekhnika, 1975 (in Russian).
81. HORI, Y., SHIMADA, S. and KASHIWABARA, H., Polymer, 20 (1979), 406.
82. RAPOPORT, N. YA., GONIASHVILI, A. SH., AKUTIN, M. S., SHIBRYAEVA, L. S., PONOMAREVA, E. L. and MILLER, V. B., Vysokomol. Soed., 23A (1981), 393.
83. CHIEN, J. C. W. and YABLONER, H., J. Polym. Sci., A1, 6 (1969), 398.
84. ZOLOTOVA, N. V. and DENISOV, E. T., J. Polym. Sci., A1, 9 (1971), 3311.
85. SHILOV, K. B. and DENISOV, E. T., Vysokomol. Soed., 19A (1977), 1435.
86. LEBEDEV, YA. S., Kinetika i Kataliz, 8 (1967), 245.
87. MIKHAILOV, A. I., BOLSHAKOV, A. I., LEBEDEV, YA. S. and GOLDANSKY, V. I., Fiz. Tverd. Tela, 14 (1972), 1172.
88. RAPOPORT, N. YA., GONIASHVILI, A. SH., AKUTIN, M. S. and MILLER, V. B., Vysokomol. Soed., 20A (1978), 1432.
89. SHANINA, E. L., ROGINSKY, V. A. and MILLER, V. B., Vysokomol. Soed., 18A (1976), 1160.

90. KLINSHPONT, E. R. and MILINCHUK, V. K., *Vysokomol. Soed.*, **18B** (1975), 358.
91. EDA, B., NUNOME, K. and IWASAKI, M., *J. Polym. Sci.*, **13** (1969), 91.
92. KAUSH, H. H. and DE VRIES, K. G., *Int. Fracture*, **11** (1975), 727.
93. TATARENKO, L. A. and PUDOV, V. S., *Vysokomol. Soed.*, **10B** (1968), 287.
94. SHLYAPNIKOVA, I. A. and SHLYAPNIKOV, YU. A., *Vysokomol. Soed.*, **17B** (1975), 358.
95. RAPOPORT, N. YA. and MILLER, V. B., *Vysokomol. Soed.*, **19A** (1977), 1534.
96. KIRYUSHKIN, S. G. and SHLYAPNIKOV, YU. A., *Dokl. AN SSSR*, **220** (1975), 1364.
97. DENISOV, E. T., *Vysokomol. Soed.*, **19A** (1977), 1244.
98. RAPOPORT, N. YA., GONIASHVILI, A. SH., AKUTIN, M. S. and MILLER, V. B., *Vysokomol. Soed.*, **20A** (1978), 1652.
99. ROGINSKY, V. A. and MILLER, V. B., *Dokl. AN SSSR*, **215** (1974), 1164.
100. ROGINSKY, V. A., SHANINA, E. L., YARKOV, S. P. and MILLER, V. B., *Vysokomol. Soed.*, **24A** (1982), 1241.
101. BRESLER, S. E., KAZBEKOV, E. I., FOMICHEV, F. N., SETCH, F. and SMEITAK, P., *Fiz. Tverd. Tela*, **5** (1963), 675.
102. BUTYAGIN, P. YU., *Dokl. AN SSSR*, **140** (1961), 145.
103. KIRYUKHIN, V. P., KLINSHPONT, E. R. and MILINCHUK, V. K., *Vysokomol. Soed.*, **18A** (1976), 1465.
104. KAUSH, H. H. and BECHT, J., *Deformation and Fracture of High Polymers*, eds J. A. Hassel and R. I. Jaffee, New York, Plenum Press, 1974, p. 317.
105. RADTSIG, V. A. and RAINOV, M. M., *Vysokomol. Soed.*, **180** (1976), 2022; RADTSIG, V. A., *Theses of Reports at 5th All-Union Symposium on Mechanoemission and Mechanochemistry of Solids*, Tallin, 1975, p. 101 (in Russian).
106. LISITSIN, A. P., EFREMOV, V. A. and MIKHAILOV, N. V., *Vysokomol. Soed.*, **16B** (1971), 123.
107. RAPOPORT, N. YA., GONIASHVILI, A. SH., AKUTIN, M. S. and MILLER, V. B., *Vysokomol. Soed.*, **21A** (1979), 2071.
108. WINSLOW, F. N., HELLMAN, M. Y., MATRYEK, W. and STILLS, S. M., *Polym. Eng. Sci.*, **6** (1966), 1.
109. YAMADA, K., KAMEZAWA, M. and TAKAYANAGI, M., *J. Appl. Polym. Sci.*, **26** (1981), 49.
110. RAPOPORT, N. YA., SHIBRYAEVA, L. S. and MILLER, V. B., *Vysokomol. Soed.*, **25A** (1983), 831.
111. FAULKNER, D. G., *Polym. Eng. Sci.*, **22** (1982), 466.
112. EMANUEL, N. M., DENISOV, E. T. and MAIZUS, Z. K., *Chain Reactions of Hydrocarbon Oxidation in Liquid Phase*, Moscow, Nauka (in Russian).
113. SHANINA, E. L., ROGINSKY, V. A. and MILLER, V. B., *Vysokomol. Soed.*, **21B** (1979), 892.
114. ROGINSKY, V. A., SHANINA, E. L. and MILLER, V. B., *Vysokomol. Soed.*, **20A** (1978), 265.
115. SHANINA, E. L., ROGINSKY, V. A. and MILLER, V. B., *Vysokomol. Soed.*, **20B** (1978), 145.
116. RAPOPORT, N. YA., LIVANOVA, N. M., GRIGOR'EV, A. G. and ZAIKOV, G. E., *Vysokomol. Soed.*, **25A** (1983), 2188.

117. RAPOPORT, N. YA. and SHLYAPNIKOV, YU. A., *Vysokomol. Soed.*, **17A** (1975), 738.
118. STIVALA, S. S., REICH, L. and KELLEHER, P. G., *Macromol. Chem.*, **59** (1963), 28; REICH, L. and STIVALA, S. S., *Elements of Polymer Degradation*, New York, McGraw Hill, 1971, p. 229.
119. DOBRODUMOV, A. V. and EL'YASHEVITCH, A. M., *Fiz. Tverd. Tela*, **15** (1973), 891.
120. GOTLIB, YU. A., DOBRODUMOV, A. V., EL'YASHEVITCH, A. M. and SVETLOW, YU. E., *Fiz. Tverd. Tela*, **15** (1973), 801.
121. LUSTON, J., GAL, E. and CHELP, K., *Proc. IUPAC Macro—82 (No. 28)*, p. 308.
122. PEEVA, L. and EVTIMOVA, S., *Theses of Reports at 15th Colloquium of Danubian Countries on Natural and Artificial Aging of Polymers*, Moscow, 1982, p. 54 (in Russian).
123. KUZ'MINSKY, A. S., *Vysokomol. Soed.*, **19A** (1977), 2191.
124. KUZ'MINSKY, A. S., KAVUN, S. M. and KIRPICHEV, V. P., *Physical and Chemical Principles of Production, Processing and Use of Elastomers*, Moscow, Khimiya, 1976 (in Russian).
125. KUZ'MINSKY, A. S., SEDOV, V. V. and KIRSHENSTEIN, N. I., *Kauchuk i Rezina*, **5** (1975), 40.
126. POPOV, A. A., RAZUMOVSKY, S. D. *et al.*, *Dokl. AN SSSR* (1975), 282.
127. PARFENOV, V. M., POPOV, A. A., KRASHENINNIKOVA, G. A. and ZAIKOV, G. E., *Dokl. AN SSSR*, **218** (1980), 396.
128. IRING, M., LASZLÓ-HEDVIG, S., KELEN, T., TÜDŐS, F., FUZES, L., SAMAY, G. and BODOR, G., *J. Polym. Sci., Polym. Symp.*, **57** (1977), 55.
129. RAPOPORT, N. YA., SHIBRIAEVA, L. S. and ZAIKOV, G. E., *Vysokomol. Soed.*, in press.

INDEX